Manual on integrated soil management and conservation practices

FAO LAND AND WATER BULLETIN

8

Manual based on the training course:
Soil Management and Conservation – Efficient Tillage Methods for Soil Conservation
held at IITA, Ibadan, Nigeria
21 April - 1 May 1997

Organized by the
Land and Plant Nutrition Management Service
of the Land and Water Development Division
and the
Agricultural Engineering Branch of the
Agricultural Support Systems Division
FAO

in cooperation with the
International Institute of Tropical Agriculture (IITA)
Ibadan, Nigeria

INTERNATIONAL INSTITUTE OF TROPICAL AGRICULTURE

Food and Agriculture Organization of the United Nations

Rome, 2000

ISBN 92-5-104417-1

Foreword

The processes of land erosion in various regions of Latin America and Africa have their origin in social, economic and cultural factors that translate into the over-exploitation of the natural resources and the application of inadequate practices for the management of soils and water. The consequences of this are damage to much of the agricultural land, with detrimental effects on food production for the growing population in these continents.

Over the last few decades, many efforts have been made to stop the degradation of agricultural land but the process of adoption of new conservationist technologies by the farmers is still slow. Furthermore, the availability of technical personnel trained for this change is limited.

The technological strategies that have been developed for the management and conservation of soil and water sometimes are not adapted for the beneficiaries, because they could not participate in the processes of the diagnosis, planning and execution of the actions. In addition, the promotion of conservation tillage systems and practices that were not adapted to specific regional requirements has created credibility problems with the farmers probably since they had been developed in other places and introduced without a correct diagnosis of the local situation.

The development of technologies that guarantee the maintenance of agricultural land productivity in Latin America and Africa is a challenge that both technicians and farmers must face through collaborative research and field work in the farmers' own environments and conditions. This includes identification of the problems of management and conservation of soils and water and a greater emphasis on the evaluation of the potential for systems of conservation tillage adapted to the specific conditions of each region.

This Manual has been put together with the objective of assisting actions by the diverse groups of human beings who intervene in the conservation of the natural resources, particularly soil and water resources and in the context of each continent, country, region or zone. The Manual brings together a collection of concepts, experiences and practical suggestions that can be of immediate use for identifying problems and for formulating, executing and evaluating actions so as to benefit and to improve the productivity and conservation of soil and water resources.

This Manual is based on the Training Course for Soil Management and Conservation, focused particularly on efficient tillage methods for soil conservation, held at the International Institute of Tropical Agriculture (IITA) in Ibadan, Nigeria from 21 April to 1 May 1997. It was jointly organized by IITA and FAO with the participation of specialists from both national and international organizations.

The publication serves as a guide that will allow technicians and farmers to jointly discover ways to solve the problems and the limitations posed by land degradation in Latin America and Africa. Participatory action by technicians and farmers will be the basis for success in benefiting these regions. It is hoped that the Manual will help to attain the ultimate objective, which is to improve the productivity of the soils and water in a rapid and efficient manner.

Acknowledgements

The present Manual is based on the Training Course on Soil Management and Conservation with Special Emphasis on Conservation Effective Tillage Methods jointly organized by José Benites, Land and Plant Nutrition Management Service (AGLL) and Theodor Friedrich, Agricultural Engineering Branch (AGSE), FAO and the International Institute of Tropical Agriculture (IITA) in Ibadan, Nigeria. The training course would not have been possible without the active support of the Director General of IITA, Lucas Brader, as well as his staff, in particular R. Booth, J. Gulley, R. Zachmann, R. Carsky, Y. Osinubi, B. Akisinde, G. Kirchof and G. Tian and we would like to express our gratitude for this support. We would further like to thank the authors of the different papers for their collaboration in this publication, particularly Elvio Giasson, Leandro do Prado Wildner, José Barbosa dos Anjos, Valdemar Hercilio de Freitas and Richard Barber, as well as Cadmo Rosell, John Ashburner, Robert Brinkman and R. Dudal for assisting with the editing of the different language versions.

Special thanks are due to Lynette Chalk for her efficient preparation of the text and formatting of this document and Riccardo Libori for his elaboration of the graphics.

Table of contents

Page

List of tables

Page

Page

List of figures

Chapter 1

Introduction

One of the main causes of soil degradation identified in various parts of Africa by the Food and Agriculture Organization of the United Nations (FAO) is the practice of inappropriate methods of soil preparation and tillage. This entails a rapid physical, chemical and biological deterioration of the soils and consequent declines in agricultural productivity and deterioration of the environment.

The natural resources and the environment of the affected areas can be significantly improved in the short-term. Compensatory measures can include the use of selected tillage methods together with complementary soil management and conservation techniques. Together these can contribute, not only towards good seedbed preparation, but also towards removing and eliminating certain limitations that affect soil productivity such as compaction, soil capping, insufficient infiltration or poor drainage and consequent unfavourable soil moisture conditions, and extreme soil temperatures.

It is unfortunate that the development of applied research concerning soil tillage and associated soil management and conservation practices focused on combating the serious and accelerated soil degradation processes occurring in Africa, has been severely limited by a lack of both professional and technical staff trained in soil conservation techniques. It has also been limited by a lack of effective policies and strategies for long term sustainable rural and agricultural development.

In the light of these observations, FAO launched a Conservation Tillage Network in 1986 so as to support national research institutions in various African and Latin American countries. The objective was both to generate technology and to spread knowledge and information concerning methods for the identification of problems related to soil management and conservation methods, and also to evaluate the potential advantages of soil conservation tillage systems.

The training course entitled "Soil Management and Conservation: Efficient Tillage Methods for Soil Conservation" was jointly organized by FAO and the International Institute for Tropical Agriculture (IITA) in Ibadan, Nigeria. The course was financed by and received technical support from the Regular Programme of the Land and Plant Nutrition Management Service (AGLL), Land and Water Development Division, together with the Agricultural Engineering Branch (AGSE), Agricultural Support Service Division of FAO.

José Benites
Food and Agriculture Organization of the United Nations (FAO)
Rome, Italy

The programme was developed through the FAO Programme for Technical Co-operation between Developing Countries (TCDC). It was assisted by the participation of experts from the Brazilian Enterprise for Agricultural Research and Rural Extension from Santa Catarina S.A. (EPAGRI), from the Soils Department of the Faculty of Agriculture, Federal University of Río Grande do Sul (UFRGS) and from the Brazilian Enterprise for Agricultural and Livestock Research (EMBRAPA). Technical experts from IITA and FAO also participated in the programme.

The course was held at the Headquarters of the International Institute for Tropical Agriculture at Ibadan, Nigeria from 21 April to 1 May 1997.

OBJECTIVES

The course was arranged to offer training for Spanish and Portuguese speaking technicians from African countries. The objective was to indicate some of the problems of soil and water conservation, to prepare strategies and plans and to organize action programmes that take account of integrated planning for soil management.

More specifically, it was hoped that at the end of the course, the participants would be familiar with a whole series of concepts, techniques and practices These included general concepts of soil management, the characteristics of soil and water degradation problems, the formulation of an efficient management strategy for soil conservation, and the principal tillage methods together with their advantages and limitations. The course also covered principles and supporting practices for conservation tillage systems, concepts and procedural methods for selecting tillage methods and participatory planning methods, with farmers to improve the efficiency of conservation tillage methods. The programme was complemented with coverage of other soil management practices, the use of soil management plans, integrated soil management programmes and the determination of priorities and incentives for financial and credit schemes.

STRUCTURE AND CONTENT

The course programme was organized under 23 themes grouped within nine modules:

General concepts of soil and water management

- Properties, processes and behaviour of the soil
- Characteristics, important factors concerning the land.

Characteristics of soil and water degradation problems

- Symptoms of the problems
- Analysis of the main causes
- Problem priorities.

Formulation of an efficient management strategy for soil conservation

- General principles for the development of soil management strategies.

Soil tillage

- Concept and objectives of tillage
- Tillage implements
- Inter-relationships between soil tillage and the physical characteristics of soil
- Principal tillage methods and their efficiency at different levels of technology.

Other technologies for soil improvement

- Soil use as determined by its agricultural capability
- Vegetative cover and contour farming
- Green manure
- Use of fertilizer and other corrective measures
- Agricultural and livestock farming
- Physical barriers placed transversely across slopes and the capture of runoff water
- Irrigation and rainwater harvesting
- Agro-forestry conservation systems
- Control of weeds, insects and diseases under systems of conservation tillage.

Planning improved soil productivity

- Participatory planning with farmers for improving methods of conservation tillage and other soil management practices
- Selection of alternative techniques.

Execution of the soil management plans

- Participatory execution of action plans.

Examples of integrated soil management programmes in Brazil

- Humid and sub-humid hilly land
- Semi-arid regions
- Acidic Savannah regions

Examples of integrated soil management programmes in Africa

- Angola
- Cape Verde
- Equatorial Guinea
- Guinea-Bissau
- Mozambique
- Sao Tome and Principe.

Class presentations were complemented with work in groups and a field visit, during which the participants had the opportunity to observe demonstration plots showing results of some of the tillage systems that had been described, and to discuss themes covered in the course modules with technical personnel from IITA. In addition, they visited the fields of some of the farmers who were managing the demonstration plots for soil management and conservation practices.

Table 1 shows a schematic diagram of the sequence and the distribution of each theme throughout the Course.

TABLE 1
Training course on "Methods of soil management and conservation: Efficient tillage methods for soil conservation"

		Monday	Tuesday	Wednesday	Thursday	Friday	Saturday	Sunday
		(21 April)	(22)	(23)	(24)	(25)	(26)	(27)
FIRST WEEK	MORNING	• Registration • Opening • Introduction • Soil properties, processes and functions	• General principles for the development of soil management strategies	• Tillage concepts and objectives • Tillage implements	• Land capability • Vegetative cover and contour sowing • Green manure • Fertilization/ corrective measures • Mixed farming • Physical barriers	• Participatory planning by farmers for soil management	• Visit to the IITA soil management experimental plots	Free
	AFTERNOON	• Characteristics of soil degradation problems	• General principles for the development of soil management strategies	• Interrelationships between tillage and soil characteristics • Principal tillage methods	• Irrigation, rain water capture • Systems of agro-forestry • Pest control	• Selection of alternative techniques	• Visit to farmers' fields	Free
		(28)	(29)	(30)	(1 May)			
SECOND WEEK	MORNING	• Participatory execution of the plans	• Humid and sub-humid hilly land • Semi-arid regions • Acidic Savannah regions	• Summary, evaluation and conclusions • Recommendations and follow-up • Closing ceremony	• Departure from Lagos			
	AFTERNOON	• Participatory execution of the plans	• Angola • Cape Verde • Equatorial Guinea • Guinea Bissau • Mozambique • Sao Tome and Principe	• Departure from Ibadan for Lagos				

Chapter 2

Key environmental and soil factors influencing productivity and management

A series of important land characteristics must be observed when assessing a particular area regarding its agricultural suitability and the need for any specific soil management and recovery practices. Apart from the environmental characteristics such as rainfall, aspects related to the topography and the actual soil conditions, the presence of any limiting factors should be examined to be able to consider the implications of adopting certain agricultural practices on the land.

TOPOGRAPHY

Topography is characterized by the slope angles and the length and shape of the slopes. Topography is an important determining factor for soil erosion, for erosion control practices and for the possibilities for mechanized soil tillage, and has a major influence on the agricultural suitability of the land.

The greater the slope angle of the land and the length of the slopes, the more severe is the soil erosion that may occur. Increased slope angles cause increased runoff velocity and with this, the kinetic energy of the water causes more erosion. Long slopes allow the runoff to build up, increasing its volume and causing yet more serious erosion.

Apart from the erosion problem, areas with steeper slopes also show less potential for agricultural use. This is due to the greater difficulty, or even the impossibility of mechanical soil tillage or transport in and from the field on steep slopes. Tillage may be further hampered by the often shallow soil depth on steep slopes.

RAINFALL

Rainfall is one of the most important climatic factors influencing soil erosion. Runoff volume and velocity depend on the intensity, duration and frequency of the rainfall. Of these factors, the intensity is the most important. Erosion losses increase with higher rainfall intensities. The duration of the rainfall is a complementary factor.

E. Giasson
Soils Department of the Federal University of Rio Grande do Sul
Porto Alegre, Brazil

The rainfall frequency also influences the soil erosion losses. When rain occurs at short intervals, the soil moisture remains high and the runoff is more voluminous, even if the rain is less intense. After longer intervals, the soil is drier and there should not be any runoff for low intensity showers, but in cases of drought, the vegetation can suffer due to the lack of moisture and so reduce the natural protection of the land.

During a heavy storm, tens of rain droplets hit each square centimetre of land, detaching particles from the soil mass. Particles can be thrown more than 60 cm high and over 1.5 metres in distance. If the land is without a vegetative cover, the droplets can break away many tons of soil particles per hectare that are then readily transported by the surface runoff.

The droplets contribute to erosion in various ways:

- they loosen and break off the soil particles in the place which suffers their impact;
- they transport the detached particles;
- they provide energy in the form of turbulence to the water on the surface.

To prevent erosion, it is necessary to avoid the soil particles being loosened by the impact of the droplets as they strike the land.

According to Wischmeier and Smith (1978), when all the other factors are maintained constant apart from rainfall, the soil loss per unit area of bare soil is directly proportional to the product of two rainfall characteristics, the kinetic energy and the maximum intensity over a period of 30 minutes. This product is used to express the potential erosivity of the rain.

SOIL LIMITATIONS

Acidity

Soil acidity depends on the parent material of the soil, its age and landform and the present and past climates. It can be modified by soil management.

Soil acidity is associated with several soil characteristics (Rowell, 1994):

- low exchangeable calcium and magnesium and low base saturation percentage;
- a high proportion of exchangeable aluminium;
- a lower cation exchange capacity than in similar, less acid soils due to reduced negative charges on the surfaces of the organic matter and increased positive charges on the surface of the oxides;
- changes in nutrient availability; for example, the solubility of phosphorus is reduced;
- increased solubility of toxic substances; for example, aluminium and manganese;
- reduced activity of many soil organisms, in extreme cases resulting in an accumulation of organic matter, reduced mineralization and low availability of nitrogen, phosphorus and sulphur.

Alkalinity and sodicity

Alkaline or sodic soil areas occur predominantly in arid regions and their appearance depends on the type of original soil material, the vegetation, the hydrology and the soil management, particularly in areas of poorly managed irrigation systems.

Soil alkalinity (pH>7) presents itself in soils where the material is calcareous or dolomitic. Sodic soils occur where there has been an accumulation of exchangeable sodium, naturally or under irrigation. Such soils have high concentrations of OH^- ions associated with high contents of bicarbonates and carbonates. Sodic soils have low structure stability because of the high exchangeable sodium and many have a dense, virtually impervious topsoil or subsoil.

Alkaline and especially, sodic conditions cause various plant nutritional problems such as chlorosis, which is caused by the incapacity of the plants to take up sufficient iron or manganese. Deficiencies of copper and zinc may occur as well, and also of phosphorus (due to its low solubility). If the soil has a high $CaCO_3$ content, potassium deficiency can occur because this can be readily leached. Nitrogen may be deficient as well due to the generally low organic matter content (Rowell, 1994).

Salinity

Saline soils have high contents of different types of salts and may have a high proportion of exchangeable sodium. Strongly saline soils may have surface efflorescence or crusts of gypsum ($CaSO_4$), common salt (NaCl), sodium carbonate (Na_2CO_3) and others.

Soil salinity can arise due to saline parent material, seawater flooding, wind-borne salts or irrigation with saline water. However, the majority of saline soils are formed through capillary rise and evaporation of water which accumulates salt over time.

Salts affect the crops through specific toxic ions, through nutrient imbalance inducing deficiencies, and through an increase in the osmotic pressure of the soil solution which causes a moisture deficiency. The soil structure and permeability may be damaged by the high exchangeable sodium remaining when the salts are leached out, unless remedial or preventive measures are taken, such as gypsum application.

Low cation exchange capacity (CEC)

The soil CEC is a measure of the quantity of negative charges present on the mineral and organic surfaces of the soil and represents the quantity of cations that can be held on these surfaces. A high soil CEC allows the soil to retain a large amount of nutrients and at the same time, keep them available for the plants.

Soils with a low CEC can only hold a small quantity of nutrients on the exchange sites. The nutrients applied to the soil that exceed this amount can easily be leached out by excess rain or irrigation water. This implies that these low CEC soils need different management as regards fertilizer application, small doses of nutrients needing to be applied frequently.

Phosphorus fixation

The fixing of phosphorus by the soil is a natural process that occurs within the soil. It can lead to phosphorus deficiency, even though the total P content might be high. Phosphate fixation is a specific adsorption process, which mainly occurs in soils with high contents of iron (hematite, goethite) and aluminium oxides (gibbsite) and of clay minerals (mainly kaolinite). These are soils typical of tropical and subtropical regions. At a low pH these soils tend to fix phosphate. By raising the soil pH through the application of lime and organic matter, the specific adsorption of phosphate is reduced.

Cracking and swelling properties

Cracking properties commonly occur in clayey soils that predominantly contain swelling clay minerals, such as those from the smectite group. These soils undergo considerable movement during expansion and contraction, because of the pronounced volume changes with variations in the soil moisture content. The soils contract and wide cracks are formed when they dry out, and they expand, turning very plastic and sticky when they are wet. This soil movement may cause the formation of a typical micro-relief on the surface (small undulations) and of wedge-shaped aggregates in the subsoil.

These soils present serious problems for tillage as they have an inappropriate consistency for this, not only when they are dry but also when they are wet. The very hard consistency when dry makes tillage extremely difficult, requiring additional tractor power, causing greater wear to the implements and not allowing the formation of a good seedbed as the clods cannot be broken down. In contrast, when they are wet they are extremely plastic and sticky, again making tillage difficult as the soil sticks to the equipment and also impedes traction and the passage of the machinery.

SOIL CONDITIONS

Depth

Soil depth can vary from a few centimetres down to several metres. Crop roots use the soil to depths ranging from a few centimetres to more than a metre; some crops have roots to a depth of many metres.

Inadequate soil depth limits root development and the availability of moisture and nutrients for the plants, and may affect infiltration and tillage options. The shallower the soil, the more limited are the types of use to which it may be put and the more limited the crop development. Shallow soils have less volume available for the retention of moisture and nutrients, they can impede any tillage or make it difficult and also, they may be susceptible to erosion because water infiltration may be restricted by the rock substrata. These adverse factors vary in severity according to the nature of the interface between the soil and the bedrock. If the soil is in contact with partially decomposed bedrock, there can be some water infiltration and root penetration and the tillage implements may be able to break it up. Solid and hard bedrock constitutes a stronger limitation.

Soil texture

The soil solid phase is made up predominantly of particles that are mineral in nature and which, according to their diameter, can be classified according to the size fractions of sand, silt and clay, in addition to coarse, medium and fine gravel.

The relative proportion of sand, silt and clay fractions making up the soil mass is called the soil texture. Texture is intimately related to the mineral composition, the specific surface area and the soil pore space. It affects practically all of the factors governing plant growth. Soil texture influences the movement and availability of soil moisture, aeration, nutrient availability and the resistance of the soil to root penetration. It also influences physical properties related to the soil's susceptibility to soil degradation, such as aggregate stability.

Consistency

A dry clod of clay soil is normally hard and resistant to fracture. As water is added to the clod and it becomes more moist, its resistance to breakage is reduced. With more water, instead of fracturing, it tends to form a lump when compressed and becomes malleable, plastic. With still more water it tends to stick to the hands.

This resistance of the soil to break-up, its plasticity and its tendency to stick to other objects are aspects of soil consistency, depending on soil texture, organic matter content, soil mineralogy and moisture content.

Determination of soil consistency helps to identify the optimum range of soil moisture contents for tillage. Under ideal conditions, the soil should not suffer compaction, the soil should not be plastic and it should be easy to prepare as it will no longer be highly resistant.

Structure and porosity

Soil structure and porosity exercise an influence on the supply of water and air to the roots, on the availability of nutrients, on the penetration and development of the roots and on the development of the soil micro-fauna. A structure of good quality implies a good quality of pore space, with good continuity and stability of pores and a good distribution of the pore size, including both macro- and micro-pores (Cabeda, 1984).

Moisture is held in the micro-pores. Water moves in the macro-pores and these tend to be occupied by the air constituting the soil atmosphere. Soil pore space is a dynamic property and changes with tillage. The limits between which its value can vary are very wide and depend upon the compaction, the nature of the soil particles and the texture of the soil. Total porosity is also intimately related to soil structure and it increases as the soil forms aggregates. Whichever agricultural practice alters soil structure, will also affect soil porosity.

According to Larson (1964), the topsoil aggregates around the seed and the young plants should be of small size in order to promote an adequate moisture regime and perfect contact between the soil, the seed and the roots. However, they should not be so small that they favour the formation of surface crusts and compacted caps. According to Kohnke (1968), the ideal size for the aggregates is a diameter between 0.5 and 2 mm. Larger aggregates restrict the volume of soil explored by the roots and smaller aggregates give rise to very small pores that will not drain but remain water-saturated.

In the deeper horizons, it is important that the structure maintain its original characteristics. One may verify whether or not there has been any structural alteration by taking a sample of moist soil and separating its aggregates. The existence of surfaces of separation between the aggregates that are angular or smooth and well defined in shape indicates structural alteration (or swelling and shrinking in certain clay soils). Irregular surfaces and the presence of tubular pores of various sizes indicate that the structure and the pore space are in a favourable condition for agricultural crop growth. The formation of this type of structure and porosity can be encouraged through management practices such as using green manure, leaving a mulch of crop residues and incorporating crops with dense root systems in the cropping sequence.

Soil density

Soil bulk density is the ratio of the mass of the dry soil particles to the combined volume of the particles and the pores. It is expressed in g/cm^3 or ton/m^3.

Soil density is related to other soil characteristics. For instance, sandy soils with low porosity have greater density (1.2 to 1.8 g/cm^3) than clay soils (1.0 to 1.6 g/cm^3) which have a greater volume of pore space. Organic matter tends to reduce soil, bulk density due to its own low density and due to its stabilizing effect on soil structure, resulting in greater porosity. Compaction resulting from inappropriate use of agricultural equipment, heavy or frequent traffic or poor soil management can raise the density of the surface horizons to values that may reach 2 g/cm^3. Soil density is often used as an indicator of compaction.

Nutrient content

Nutrient availability is fundamental for crop development. The soil nutrient content depends on the parent material and the soil formation process (the original soil content), on the supply and nature of fertilizer, on the intensity of leaching and of erosion, on the absorption of the nutrients by the crops and on the CEC of the soil.

Although nutrient deficiency may be easily corrected in many cases, soils with better natural availability of nutrients will require less investment and thus, show a natural aptitude for better yields. Knowledge of the need or not for applying large quantities of nutrients in the form of fertilizer as compared to the availability of resources, is a determining factor for the recommended type of land use.

In addition to evaluating the contents and proportions of exchangeable cations (Ca^{++}, Mg^{++}, K^+ and Na^+), it is also necessary to evaluate the soil nitrogen content (through the organic matter), its available phosphorus content, its content of essential micro-nutrients and the value of the CEC of the soil.

Soil organic matter and soil organisms

The soil organic matter is made up of all the dead organic material of animal or vegetable origin, together with the organic products produced by its transformation. A small fraction of the organic matter includes original materials only slightly transformed; another fraction, products that are completely transformed, dark in colour, of high molecular weight and which are called the humus compounds (humic and fulvic acids, humin).

After fresh organic residues are added to the soil, there is a rapid rise in the organism population due to the abundance of easily decomposed material, including sugars and proteins. These are transformed into energy, CO_2 and H_2O and into compounds synthesized by the organisms. As the amount of easily decomposable organic matter declines, the number of organisms also diminishes. The successors to these then attack the remaining, more resistant compounds of cellulose and lignin and also the synthesized compounds, their overall proportion gradually reducing as the amount of humus increases. The speed of transformation of the fresh organic residues to humus depends on the nature of the organic matter supplied and the environmental conditions in the soil.

After application of, for example, woody materials or other organic residues that have a high carbon and low nitrogen content (a high C/N ratio), the organisms consume the N available in the soil, immobilizing it. As a result, for some time there will be very little N available to the plants. With the gradual decomposition of the organic matter, the population of organisms reduces and the N once again becomes available for the plants, establishing a C/N ratio of between 10 and 12. In order to avoid competition between the organisms and the plants for the N, it is convenient to wait until the organic residues reach an advanced stage of decomposition before establishing a new crop.

Organic matter added to the soil normally includes leaves, roots, crop residues and corrective organic compounds. As most of the vegetative residues are applied to the surface or the topsoil, the organic matter content in this layer tends to be higher and to decrease with depth.

The nutrient content of the organic matter is important for the plants. Through the activity of the flora and fauna present in the soil. These nutrients are transformed into inorganic substances that are then available to the plants. As yields are increased, correct use of mineral fertilizer and root mass increases the organic matter content of the soil due to the greater amount of crop residues and root mass that will be incorporated. Organic matter can also be added through the use of green manure and by applying organic residues (manure, compost).

Organic matter favours the formation of a stable structure in the soil through a close association of clays with the organic matter. It increases the water holding capacity as it can absorb water to a ratio of three to five times its own weight, which is very important in the case of sandy soils. Organic matter increases the retention of soil nutrients in a form available to the plant due to its capacity to exchange cations (the CEC of humus ranges from some 1 to 5 meq/g).

The soil fauna, especially earthworms, create vertical macropores of various sizes in undisturbed soil, increasing aeration, infiltration rate and permeability. The soil microflora produces glue-like substances, including polysaccharides, that help stabilize the soil structure.

Tillage affects the physical characteristics of the soil and may increase the porosity and aeration, but also negatively affects the soil fauna due to the soil disturbance caused by the agricultural implements. Minimum tillage and zero tillage systems safeguard the soil fauna and the pore structure created by them. Because these systems tend to maintain more stable soil temperature and moisture regimes, they also protect the microbial population during periods of high temperature and prolonged drought. Burning of the stubble tends to reduce the micro-flora, particularly near the soil surface. Leaving the crop residues on the soil surface and using a perennial vegetative cover crop or using plants with a dense root system will favour a better development of the soil fauna and the microbial biomass.

Fertilization, whether organic or mineral, tends to stimulate the soil organisms. The use of pesticides can dramatically reduce the number of organisms. The practice of monoculture cropping can affect the population either by continuously supplying the same type of organic material or through the accumulation of toxic substances exuded by the roots, thus reducing the diversity of the species and upsetting their equilibrium.

PRODUCTIVITY

Productivity is a good indicator of the land conditions, since it directly reflects changes in the qualities and limitations of the land. Assessment of the productivity of specific target areas and comparison with similar neighbouring areas that are already used under adequate practices of crop management allows identification of the needs for applying particular soil improvement practices.

The main objective of sustainable agriculture is to achieve high productivity without degrading the soils. Productivity shows a response to all the factors that control the growth, development and production of the crops. Sustained good productivity is synonymous with good land conditions and good management practices, which at the same time maintain or improve the land qualities.

Chapter 3

General principles for the development of soil management strategies

OBJECTIVES OF SOIL MANAGEMENT FOR AGRICULTURE

The main objective of soil management for agriculture is to create favourable conditions for good crop growth, seed germination, emergence of the young plants, root growth, plant development, grain formation and harvest.

Desirable edaphological conditions are as follows:

- physical conditions that favour seed germination (aggregate size, moisture content and soil temperature). The optimum size of the aggregates varies with the seed size and must be such that maximum contact can be achieved between the seed and the soil so as to facilitate the movement of soil moisture to the seeds without oxygen deficiency. Plant germination is seriously limited by either an excess or a shortage of soil moisture and also by extreme temperatures;

- a surface structure that does not impede emergence of the young plants. The presence of hard crusts restricts emergence. There are also interactions between the strength, thickness, compaction and moisture content of the crust and the crop type, seed depth, size, and vigour;

- a soil structure, porosity and consistency in the uppermost horizon that favours the initial growth of the young plant and its roots. Early growth is retarded by clay soils with large and hard aggregates and by sandy soils that become massive and hard when they dry ("hardsetting" soils);

- a structure, size and continuity of the pores in the subsoil that allows free penetration and development of the roots. The presence of hard pans due to tillage, or compacted layers due to natural compaction processes, restricts root penetration and the volume of soil that the roots can explore to absorb moisture and nutrients. In addition, hardpans weaken the anchorage of many crops;

R. Barber, Consultant
Food and Agriculture Organization of the United Nations (FAO)
Rome, Italy

- an adequate and timely supply of nutrients that coincides with crop demand during the growth cycle. The management system must maximize the recycling of nutrients within the profile and within the farm, and it must minimize nutrient losses due to natural processes and to management. The goal of the nutrient management system is that the only nutrient losses from the soil should be those which are exported from the farm with the harvest;

- the lack of toxic substances in the soil. A high degree of saturation of the Effective Cation Exchange Capacity (ECEC) with aluminium or manganese, soluble salts, or an excess of exchangeable sodium may be toxic to many crops. There is however, much variation in crop tolerance to these;

- an adequate and timely supply of moisture to the crop throughout its cycle, particularly at critical growth stages. Excessive moisture during the initial stages of plant development can be harmful for many crops and conversely, a shortage in stages more sensitive to moisture availability such as during flowering and grain formation can seriously reduce yields. Excessive moisture during harvest can reduce yields due to lodging and grain rot. Furthermore, combine harvesters working in wet soils can degrade the soil structure and porosity;

- an adequate and timely supply of oxygen to the crop roots and the aerobic soil micro-organisms. Poor drainage conditions cause a lack of oxygen in the soil due to its diffusing some 10 000 times more slowly through water as contrasted with diffusion through air. In this way, it cannot satisfy the oxygen requirements of either the roots or aerobic micro-organisms. Lack of oxygen results in physiological stress, which affects the absorption of nutrients by the plants, and in the production of toxins due to microbiological reduction processes;

- high biological activity in the soil. The diversity of the fauna and of the micro-organisms, particularly the mesofauna population, is very important for sustaining soil productivity. Worms and termites, for example, influence soil porosity, the incorporation of organic residues and the process of humus formation;

- stable conditions in the cropped area so that flooding, water erosion or high winds do not harm the crops. Floods can cause physical damage to the crops and reduce the oxygen diffusion rate within the soil. Erosion by water reduces soil fertility and can cause land loss through the development of gullies or landslips. Strong winds can cause crop damage and the loss of leaves and flowers. Furthermore, wind erosion can occur, and the winds accentuate the moisture deficit by increasing the rate of evaporation. During cold spells, the combination of low temperatures and high winds produces a wind-chill effect which gives even lower effective temperatures, causing adverse physiological processes in the crops.

PRINCIPLES FOR THE DEVELOPMENT OF SOIL MANAGEMENT PRACTICES

There are nine general principles that should be considered as basic guidelines for the development of strategies for soil management systems: increase the soil cover, increase the soil organic matter content, increase infiltration and water retention, reduce the runoff, improve the rooting conditions, improve the chemical fertility and productivity, reduce production costs, protect the field, and reduce soil and environmental pollution.

1. Increase the soil cover

This is the most important principle for sustainable soil management as it brings multiple benefits:

- ***Reduction of water and wind erosion***

Soil cover protects the surface from the force of the rain droplets and reduces the separation of the particles of soil aggregates, which is the first step in the process of erosion by water. There is evidence that a 40 percent soil cover reduces the soil losses due to splash erosion to values of less than 10 percent of what would be expected from same soil when bare (Figure 1). When soil erosion is caused by a combination of erosive processes, such as splash erosion and rill erosion, then it is likely that a cover of more than 40 percent is needed to reduce losses to only 10 percent of those incurred by the same soil in a bare condition. Studies in Kenya concerning the effect of different mulch covers on soil losses from simulated rainfall which caused both rill and splash erosion, showed that a cover of between 67 and 79 percent was needed to reduce soil losses to 10 percent of those from the same soil in a bare state (Table 2).

TABLE 2
Mulch cover and soil loss from two simulated rainfalls (Barber and Thomas, 1981)

Mulch (t/ha)	Cover (%)	Soil loss (t/ha)		
		1st Storm	2nd Storm	Average
0	0	1.40	6.27	3.84
1	46	0.22	1.70	0.96
2	67	0.12	0.83	0.48
4	79	0.03	0.26	0.15
Average		0.44	2.27	1.36

The runoff velocity and the capacity of runoff to transport loose particles both increase markedly with slope angle. Therefore, the cover that is in contact with the soil is very important, even more than the aerial cover. The cover in contact with the soil not only dissipates the energy of the raindrops but also reduces runoff velocity and consequently soil losses because of less particle transportation (Paningbatan *et al.,* 1995). Empirical studies in El Salvador have shown that a contact cover of approximately 75 percent is needed to achieve “low” erosion risks (Figure 2). This figure approximates to the 67-79 percent range of cover cited in the Kenyan study as being necessary to reduce soil losses to ten percent of those from bare soil.

The presence of a protective cover also reduces wind erosion by reducing the wind velocity over the soil surface (Table 3).

TABLE 3
Average effect of the nature and orientation of crop residues on the erosion of a sandy loam soil by wind at a uniform velocity (Finkel, 1986)

Quantity of residues on the surface (t/ha)	Quantity of soil eroded in the wind tunnel (t/ha)			
	Wheat residues		Sorghum residues	
	Stubble 25 cm high	Flattened stubble	Stubble 25 cm high	Flattened stubble
0	35.8	35.8	35.8	35.8
0.56	6.3	19.0	29.1	32.5
1.12	0.2	5.6	18.1	23.3
2.24	traces	0.2	8.7	11.9
3.36	traces	traces	3.1	4.9
6.72	traces	traces	traces	0.4

FIGURE 1
Effects of cover on the reduction of splash erosion (Source: Shaxson *et al.*, 1989)

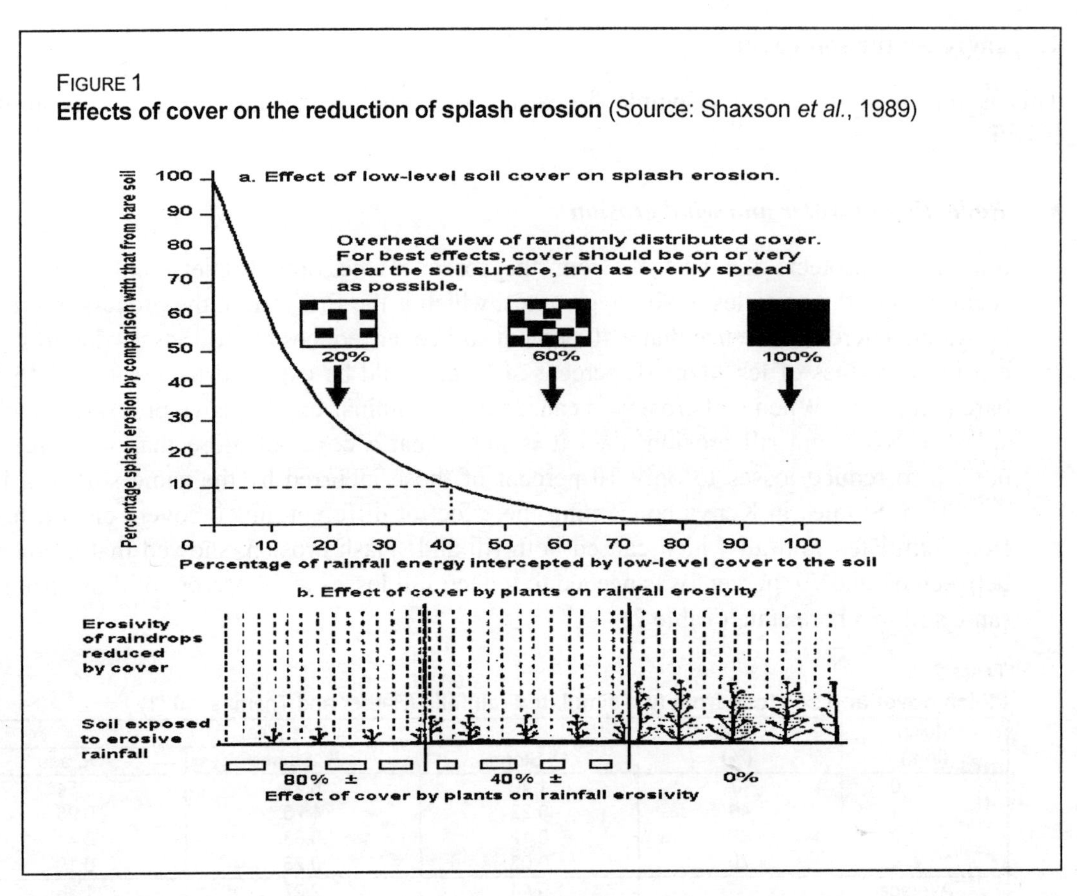

FIGURE 2
Relationship between soil cover, slope and erosion risk (Argueta, 1996)

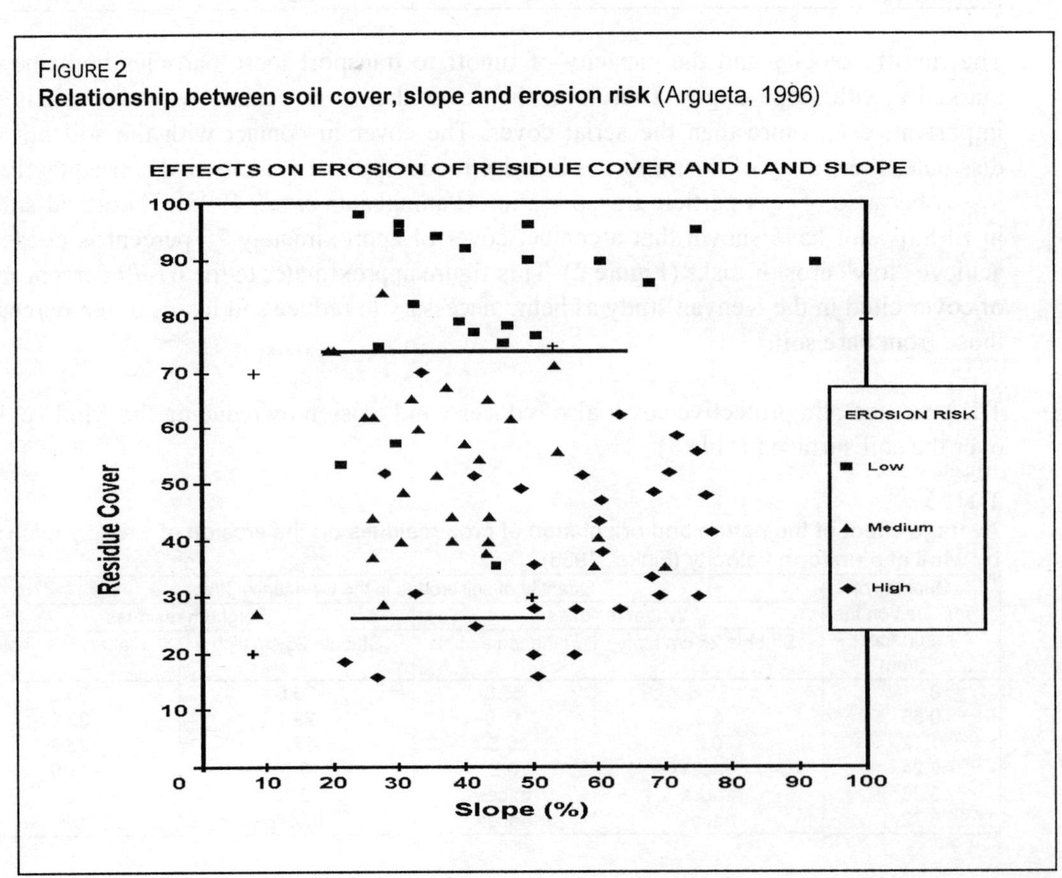

- ***Increase of the rainfall infiltration rate***

  The soil protection provided by cover prevents the formation of surface crusts and maintains a higher infiltration rate. Figure 3 shows the difference in infiltration rates for a soil in Nigeria, with and without cover (Lal, 1975).

FIGURE 3
Infiltration rates in plots with and without mulch (Lal, 1975)

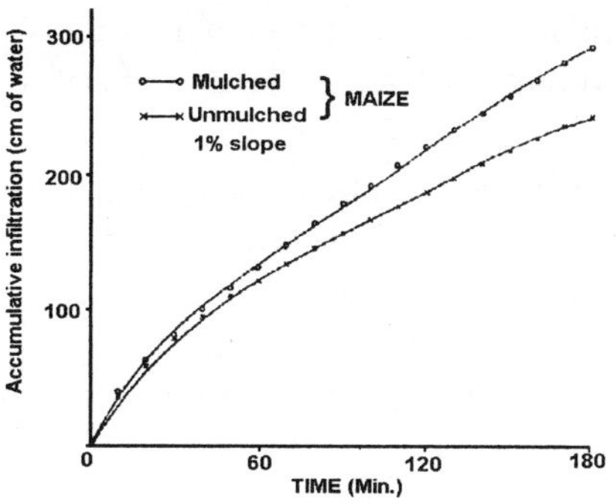

- ***Reduction of moisture loss by evaporation and increase in moisture availability***

  The combination of improved infiltration and lower moisture loss through evaporation results in increased moisture availability for the crop. Table 4 shows how the presence of a mulch increases the amount of moisture stored in the soil.

TABLE 4
Effects of mulch and the type of tillage on the amount of moisture (mm) stored in 120 cm soil depth. Faizabad, India, 1983 (Sharma, 1991)

Tillage system	Without mulch	With mulch	Average
Minimum tillage	117	154	135.5
Reduced tillage	150	181	165.5
Ploughed	165	185	175.0
Average	144	173	-
$LSD_{0,05}$	8.4	14.3	

- ***Reduction of the temperature***

  The presence of a cover substantially reduces the daily maximum temperature in the first 5 cm of soil depth. In zones and seasons where temperatures are very high, a cover will have beneficial effects on seed germination, biological activity, microbiological processes and initial crop growth. For many crops, temperatures above 40°C inhibit seed germination and temperatures higher than 28-30°C at 5 cm depth restrict the growth of seedlings of many crops (Lal, 1985).

- ***Improvement of conditions for germination***

  Increased moisture and lower temperatures create better conditions for seed germination. Table 5 shows data concerning moisture, temperature and the percentage emergence for cowpea and soybean as these relate to different tillage systems. The data compares zero tillage with conventional tillage, representing contrasting systems with and without cover.

TABLE 5
Types of tillage and their effect on the moisture, temperature and rate of emergence for cowpea and soybean. (Source: Nangju *et. al.*, 1975)

Tillage	Maximum soil temperature (°C)	Soil moisture content (%)	Emergence of young plants (%)	Days to emerge	Fresh weight of young plants (g)
Cowpea:					
Conventional	41	11.2	89.4[b]	4[b]	1.32
Zero	36	14.4	97.8[a]	3[a]	1.60
Soybean:					
Conventional	41	11.6	33.4[d]	6[d]	0.53
Zero	36	14.3	53.9[c]	5[e]	0.43

a, b, c, d = differences at 5% probability.

- ***Increase in organic matter content of the surface soil layer***

  Figure 4 shows that the increase in the accumulation of organic matter in the soil is directly related to the amount of residues applied as cover. The highest increase in the organic matter content initially occurs in the first 15 mm depth of soil under zero tillage, and as time passes, the organic matter content of the deeper layers also increases.

  Figure 5 shows the distribution of organic matter in the soil after 10 years of zero tillage where the residues have remained on the soil.

FIGURE 4
Relationship between the organic matter in the first 15 mm of soil and the quantity of crop residues applied over 5 years in Georgia (Langdale *et al.*, 1992)

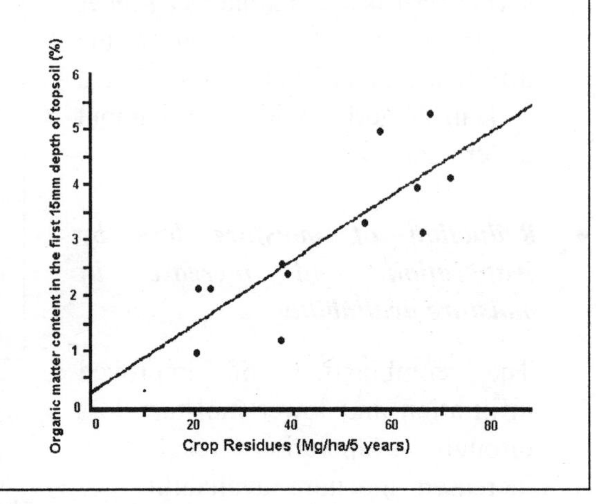

- ***Improvement of the structural stability of the surface aggregates***

  The increase in the organic matter content of the soil improves the resistance of the aggregates to erosion and to crusting.

FIGURE 5
Distribution of organic matter in the soil after 10 years of zero tillage and conventional tillage (Edwards *et al.*, 1992)

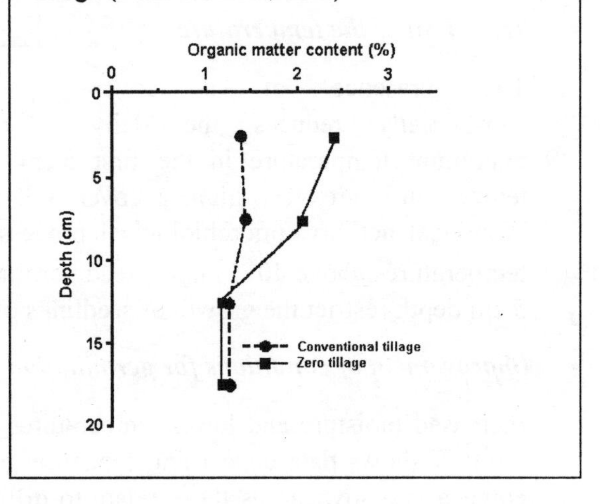

- ***Stimulation of biological activity in the soil***

  Optimum conditions of moisture and temperature stimulate the activity of the micro-organisms and of the fauna. The macro-fauna also need a dead vegetative cover on the surface for their food. Mulch has an enormous influence on the number of earthworm casts in a maize field (Table 6). In addition, Lal *et al.*, (1980) have obtained a linear relationship between earthworm activity and the quantity of mulch applied.

TABLE 6
Application of mulch and the quantity of earthworm casts (Source: Lal, 1975)

Treatment	Worm casts per m^2	Equivalent weight (t/ha)
With mulch over all the area	568	127
Mulch between rows	264	59
Without mulch	56	13

- ***Increased porosity***

  The increase in the activity of the macro-fauna results in increased soil porosity (Lal *et al.*, 1980), particularly the macro-porosity that serves as a by-pass for drainage of the major proportion of the rainfall. This leads to less leaching of soil nutrients located further from the macro-pores. Another consequence of improved porosity due to the activity of the macro-fauna, is the improved infiltration rate as shown in Table 7.

TABLE 7
Effects of cover crops on infiltration rates with and without earthworm activity (Wilson *et al.*, 1982)

Type of cover crop	Cumulative infiltration (cm/3h)		Equilibrium infiltration rate (cm/3h)	
	With worms	Without worms	With worms	Without worms
Brachiaria sp.	490	64	75	19
Centrosema sp.	220	72	30	18
Pueraria sp.	270	76	90	16
Stylosanthes sp.	390	74	60	16

- ***Stimulates biological pest control***

  Improved biological conditions stimulate the proliferation of predatory insects of the pests.

- ***Suppression of weed growth***

  In general, a good residue cover helps to suppress the emergence of many weeds. However, if the cover is insufficient, weeds can become problematic, particularly with certain species.

  Mechanisms to achieve a better cover are as follows:

  - Leave all the crop residues in the field, do not burn them, do not carry them away from the field and do not graze them, or at least reduce grazing to a minimum. This implies fencing the fields to control the grazing intensity. If farmers normally remove the residues for livestock feed, it will be necessary to review the whole farming system so as to identify how to produce alternative fodder sources to substitute for the residues. A simple model that helps to quantify the additional quantity of fodder required is shown in Figure 6.
  - Practise a system of conservation tillage that leaves the residues on the soil surface and does not bury them as in conventional tillage systems.
  - Apply organic materials as manures or mulch to increase the soil cover.
  - Increase the production of biomass in the field by sowing cover crops, intercrops, relay crops and increase the population density of the crops.
  - Sow crops that produce large quantities of residues (see Table 8) within the overall crop rotation.
  - Increase the chemical fertility of the soils through the application of fertilizers and organic manures so as to produce greater quantities of biomass.
  - Leave the dead weeds on the surface as a cover through the use of herbicides or by mechanical weed control with field cultivators that uproot the weeds and leave them on the soil surface rather than ploughing them in.
  - Leave stones on the surface in manual systems as these serve as a cover which will increase the infiltration rate of the rainfall. This is preferable to removing them for the construction of dead barriers.

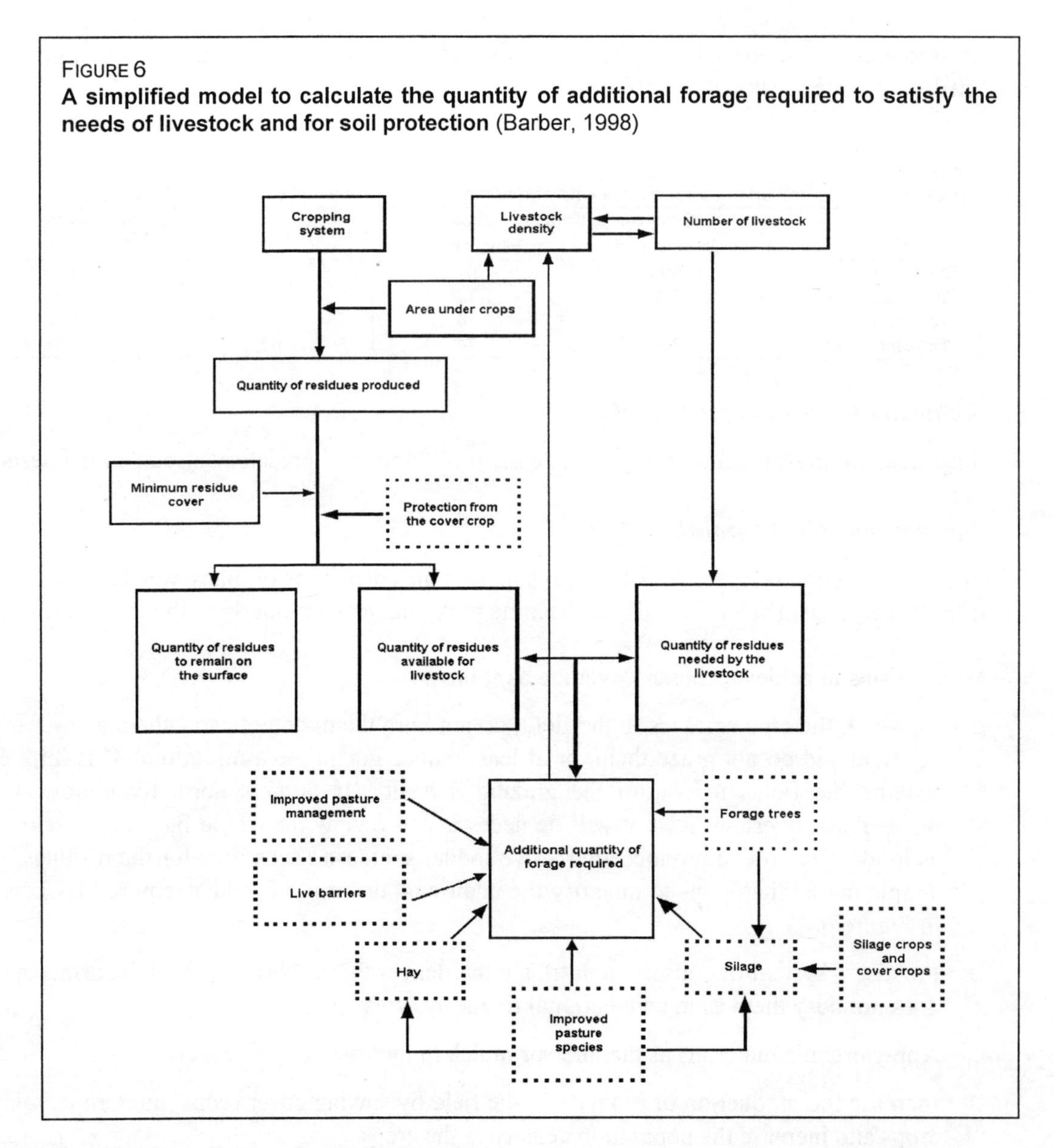

FIGURE 6
A simplified model to calculate the quantity of additional forage required to satisfy the needs of livestock and for soil protection (Barber, 1998)

2. Increase the soil organic matter content

This principle is intimately related to the preceding one related to increasing soil cover, because, on increasing soil cover with organic materials, the organic matter content of the more superficial horizons is increased. It is more difficult to increase the organic matter content of the lower horizons and particularly, those in the subsoil. The beneficial effects of an increase in soil organic matter content are as follows:

- ***Increase in the stability of surface aggregates***

  This leads to greater resistance of the aggregates to crusting, water and wind erosion, and a higher infiltration rate.

- ***Increase of the moisture retention capacity of the soil***

  The increase is often not very great, except in very sandy soils.

TABLE 8
Straw production and the relationships of C/N and of the grain/straw weight for annual crops, Santa Cruz, Bolivia (Barber, 1994)

Crop (season)	Straw kg/ha	C/N Ratio	Grain/straw weight Ratio
Summer crops			
Soybean	1 570	22	1.56
Maize	3 760	40	0.51
Sorghum for grain	3 600	32	0.82
Cotton	3,520	22	0.29
Winter crops			
Soybean	900	22	1.56
Wheat	970	75	1.70
Sorghum for grain	2,680	32	0.82
Beans	900	26	0.87
Sunflower	3,590	33	0.34
Fallow crops			
Crotalaria juncea (summer)	7,590	19	-
Avena strigosa (winter)	3,010	28	-

- ***Increase in the capacity of the soil to retain nutrients***

  This is attributed to the increase in the cation exchange capacity of the soil, and as in the case of the moisture retention capacity, the increases are often not very large, except in very sandy soils.

- ***Stimulation of the soil biological activity***

  An increased activity of the soil macro-fauna will result in more soil macro-porosity and greater incorporation and humification of the organic residues.

The principles or mechanisms for increasing soil organic matter content are the same as those for increasing soil cover, except for the practice of leaving stones on the surface.

3. Increase the water infiltration rate and moisture retention capacity

The beneficial effects of increasing infiltration rate and moisture retention capacity of soils are as follows:

- ***Reduction of crop moisture deficit***
- ***Increase of the yield and production of the crop biomass***
- ***Reduction of runoff.*** This results in a reduced loss of water, soil, fertilizer and pesticides which could otherwise contaminate the environment.

Mechanisms to increase soil infiltration rates and moisture retention capacity are as follows:

- Maintain a protective cover of residues over the soil to avoid the formation of surface crusts that would impede the infiltration of the rain. The presence of the cover protects the soil from the impact of the rain droplets and avoids degradation of the aggregates and the formation of surface crusts, so facilitating the infiltration. In addition, the contact between the residues and the soil slows down the runoff giving more time to infiltrate. For these reasons, all crop residues should be left on the soil surface and a conservationist tillage system practised that does not bury the residues.

- Reduce the moisture losses due to evaporation by reducing the wind velocity. This can be achieved by installing windbreaks.
- Create surface roughness between the crop rows to delay crust formation and thus encourage water infiltration. Tillage, usually ploughing, can achieve this, leaving large aggregates on the surface of the soil, which thus encourages better infiltration and root growth. In West Africa, ploughing is done at the end of the rainy season. Although this does not leave stubble on the surface, the surface roughness encourages infiltration. It would be even more conservation oriented if there were residues to protect the surface. Lal (1995) suggests that permanent benefits can only be achieved if the organic matter content of these soils is improved through the use of fallow crops, by controlled grazing and by not burning the residues.
- Cultivate the soil with a field cultivator after each rainfall. However, if too much tillage is done then biological degradation of the soil is encouraged and maintenance of a protective residue cover on the surface is made more difficult. Soil crusting problems arise mainly where there are no residues and in soils with high contents of fine sand. Surface compaction problems occur most commonly in hardsetting soils which are most usual in light and medium textured soils.
- Increase the time available for infiltration of the rainwater by resting the soil before establishing the crop. This works better where it is feasible to grow two crops each year and it is possible to sacrifice one of the cropping seasons. To avoid exhausting the moisture that accumulates during the rest period, it is necessary to control the growth of vegetation during the fallow period but without leaving the soil bare.
- Create small barriers that impede the runoff and give more time for water infiltration. This can be achieved by tilling and establishing the crop along the contours, which creates small undulations across the slope. Contour ridging, whether they are tied ridges or continuous ones, further increases the time available for infiltration. However, this is not advisable on slopes steeper than seven percent due to the risks of overflow and consequent erosion.
- Improve the permeability of compact layers that are impeding moisture percolation to greater depth, so increasing the moisture retention capacity of the profile. This may be achieved through deep tillage that loosens the impermeable layer and increases the porosity.
- Apply organic fertilizer to increase the moisture retention capacity of the soil. Normally, large quantities are required and there will be more beneficial effects in sandy soils with low moisture retention values.
- Reduce the slope of the land to give more time for rainwater infiltration, for example by constructing bench terraces, orchard terraces or individual terraces.

4. Reduce the runoff

The beneficial effects of reducing runoff are as follows:

- ***reduction of the losses of soil, water, nutrients, fertilizers and pesticides.*** This results in less erosion of the field and less environmental pollution of downstream waters;
- ***increase the moisture available to the crop and consequently grain yield and biomass production.***

There is a close relationship between rainwater infiltration and the commencement of runoff. It follows that the factors influencing infiltration will also influence runoff initiation. Measures to reduce runoff once it has commenced are considered below.

- Collect the runoff in structures within which the water can infiltrate. The size, number and distance between the structures must be sufficient to collect all the runoff and avoid overspill that could cause erosion. Examples include basins or pits which are more appropriate for perennial crops, and dead barriers constructed of stones which are more appropriate for mechanized or animal traction systems. Accumulating the stones in barriers avoids possible damage to tillage implements. For manual systems it is better to leave the stones on the surface to act as cover to encourage rainfall infiltration, rather than removing them and leaving the soil bare and more susceptible to erosion.

- Construct structures that collect and lead the runoff away from the field. Hillside ditches and cut-off drains, made by hand or with machinery, can collect and lead the runoff water out of the field at a reduced velocity. It is important to construct the channels with a gradient that is sufficient to carry the runoff at a velocity that does not cause erosion. There must also be a main drainage course where the runoff can be discharged but without causing erosion problems due to the increased flow at the point of entry into the drainage course or along the drainage course.

- Establish permeable barriers along the lines of contour to reduce runoff velocity, so creating conditions that favour its infiltration, such as vegetative barriers (live barriers). The type of vegetation selected will influence the efficiency of the barrier in reducing the runoff. Particularly important factors include the type of vegetation, its growth habit, its density (i.e. the degree of contact between the soil and the vegetative stems), the width of the vegetative barrier, the length and angle of the slope and the presence of surface crop residues in the field.

5. Improve the rooting conditions

The beneficial effects of improving the conditions for roots are:

- ***improved root development and growth, and as a result the absorption of moisture and nutrients by the crop;***
- ***reduced probability that the crops will suffer from drought.***

Mechanisms for improving crop rooting conditions are as follows:

- Carry out deep tillage to loosen any compacted or hardened massive layers that are impeding root penetration. Loosening these horizons will increase the porosity, so allowing the roots to penetrate. Table 9 shows the effect of deep tillage with a disc plough and subsoiler on some physical properties, and on the root development of soybean in a compacted soil in Bolivia. As far as possible, one should use conservation tillage methods that do not bury surface trash or bring large aggregates from the subsoil to the surface. For these purposes, a subsoiler is preferable and more conservationist than either a moldboard or a disc plough. A chisel plough can be used in soils in which compaction is only incipient, but in well compacted soils the subsoiler should be used.

TABLE 9
Effects of deep tillage on some physical properties and root development in a compacted soil (Barbosa *et al.*, 1989)

Tillage	Bulk Density (Mg/m³)		Porosity	Root depth	Root weight	Penetrometer resistance
	0.05-0.1m	0.15-0.2m	(%)	(m)	(g/plant)	(MPa)
Disc harrow	1.61	1.73[a]	35[a]	0.24[a]	3.7[b]	2.0[a]
Disc plough	1.54	1.56[b]	41[b]	0.32[b]	4.8[ab]	1.4[b]
Sub-soiling	1.58	1.64[ab]	38[ab]	0.33[b]	4.1[b]	1.2[b]
Sub-soiling with controlled traffic	1.49	1.55[b]	41[b]	0.31[b]	5.8[a]	1.3[b]

Numbers in a column followed by a different letter are significantly different at a 95% probability level.
Source: Orellana *et al.*, 1990.

- Improve drainage by installing drainage channels where soils are poorly or imperfectly drained, and where a lack of oxygen impedes root development. The construction of raised beds is another practice that increases the depth of rooting without drainage problems. The furrows between the beds can be made with a gentle slope to assist drainage of excess water.
- Improve the chemical conditions where there is a nutritional deficiency, a nutrient imbalance or the presence of toxic substances that inhibit root growth. The most common nutritional problems affecting root development are phosphorus deficiency and toxic levels of aluminium.

6. Improve the chemical fertility and productivity

Beneficial effects produced by improving the chemical fertility and soil productivity are:

- ***increased yields;***
- ***increased production of crop biomass.***

Increased foliage and root production of the crop will provide additional residues and hence, better soil cover and increased recycling of organic matter back to the soil.

Mechanisms to increase chemical fertility and productivity of the soils are as follows:

- Undertake a careful analysis of the nutritional state of the soil and also preferably, of the plant so as to counteract whatever nutritional deficiency or nutrient imbalance that has been identified. Foliar analysis will contribute greatly to interpretation of the nutritional state, but it is most important to sample the appropriate part and during the appropriate season to enable a correct interpretation of foliar analysis. As regards inorganic fertilizer, it is important to establish the most economic application rate, the application rate corresponding to maximum yield, and the most appropriate method and time of application.
- Take advantage of using whatever organic materials are locally available for the improvement of soil fertility, as this will have beneficial effects on the physical and chemical soil properties.
- Introduce crop rotations to increase soil productivity due to their beneficial effects in counteracting weed infestation, the incidence of disease and pests, and crop competitions for moisture and nitrogen. Crop rotations also tend to rejuvenate soils, particularly soils that are " exhausted" .

- Avoid wastage of nutrients. Do not allow burning of residues or stubble, nor the export of nutrients out of the farm, and preferably not out of the field, except for those nutrients in the harvest.

- Increase the soil organic matter content, particularly in sandy soils with generally low fertility. This may be done through the application of large amounts of organic manures and mulches, sowing legumes and cover crops, intercropping, relay cropping, crop rotations, increased plant densities and through an increase of chemical fertility so as to encourage a high production of biomass.

- Try to substitute the use of nitrogenous fertilizers by sowing legume crops as part of the rotation, as intercrops, relay crops or as cover crops.

- Take advantage of the processes of nutrient recycling, particularly in zones suffering serious leaching problems. Introduce crops with deep rooting systems that absorb nutrients from the deeper layers which are normally beyond reach of most crops. In this manner, the nutrients are brought to the surface in the dead leaves and stubble to be later used by the roots of other crops. Deep rooting crops can be introduced into crop rotations, agro-forestry systems such as alley crops, or in natural fallows or enriched fallows.

- Overcome soil toxicity problems due to high levels of aluminium or manganese. This may be achieved through the substitution of the crops or varieties with other, more resistant crops or varieties, or through the substitution of aluminium or manganese cations by calcium or magnesium from applications of lime or dolomitic lime, with or without gypsum to speed up the effect.

7. Reduce production costs

The positive effects of a reduction in the costs of production are:

- ***increase in net profitability***
- ***more sustainable production systems***

The principles and mechanisms to reduce production costs are as follows:

- Whenever possible, use biological pesticides and botanical or semi-botanical herbicides, and practise integrated pest management to reduce pesticide costs.

- Reduce the need for inorganic fertilizers by sowing legume crops that form nodules without the need for inoculants.

- When available, apply rock phosphate to replace inorganic fertilizer on soils where rock phosphate is effective.

- Apply economic doses of inorganic fertilizers, in a form and at a time during the season for maximum efficiency.

- Apply organic fertilizers when available to reduce the use of inorganic ones.

- Where labour is scarce or expensive, introduce manual planters with fertilizer hoppers to speed up sowing and fertilization.

- Take maximum advantage of those management systems that involve nutrient recycling. Use deep rooting crops as fallow crops in crop rotations and as alley crops and forage alleys in agro-forestry systems. Ensure that whenever possible, the crop residues are returned to the field and not burned or grazed.

8. Protect the field

The field should be protected against the effects of flooding, erosion by water, landslips, strong winds and wind erosion. Strong winds cause not only wind erosion problems but also problems for the timely application of herbicides and insecticides.

The principles for protection against flooding are as follows:

- Install diversion canals to capture the runoff that enters the field divert and discharge it into a drainage course, avoiding any erosion at the canal exit point or along the length of the drainage course.

The principles for protecting the field against erosion by water are as follows:

- Provide maximum soil cover, increase the infiltration rate and reduce the runoff.

The principles for field protection against landslides are as follows:

- Introduce tree crops or deep-rooted crops in association with trees. The deep roots will help to stabilize the soil and increase moisture absorption through transpiration by the trees and crops.

- Install diversion canals to reduce the entry of surface and subsurface water to the field, which would otherwise reduce soil stability and favour landslides.

The principles for protecting the plot against wind erosion and high winds are as follows:

- Install windbreaks to reduce the wind velocity, provide maximum soil cover and create a rough soil surface.

9. Reduce pollution of the soil and the environment

The principles to reduce pollution of the soil and the environment are as follows:

- Apply integrated weed and pest management measures, rather than just using pesticides. Wherever possible, replace toxic by non-toxic pesticides, preferably using biological or botanical pesticides.

- Train farmers in methods to correctly manage agrochemicals.

- Apply split applications of fertilizers according to the nutrient needs of the crop and the retention capacity of the soil, so as to avoid fertilizer losses in surface runoff or groundwater.

- Apply soil conservation practices to reduce to a minimum the amount of sediments and pesticides removed from the land.

- Monitor the quality of surface and sub-surface waters so that this can serve as baseline data for assessing the efficiency of soil management practices.

Chapter 4

Concepts and objectives of tillage in conservation farming

WHY CONSERVE THE SOIL?

Erosion

Agricultural use of land is causing serious soil losses in many places throughout the world. It is probable that the human race will not be able to feed the growing population if this loss of fertile soils due to continues at the same rate. There are several causes of inadequate land use. In many developing countries, hunger is forcing people to cultivate land that is unsuitable for agriculture and which can only be converted to agricultural use through enormous efforts and costs, such as those involved in the construction of terraces.

However, the most serious damage occurs in the large areas under mechanized agriculture as this takes place on a much larger scale. The United States of America serve as an example where during the 1930's, vast areas of fertile soil were damaged by wind erosion and temporarily abandoned. Today, the same types of mistake are still causing enormous losses of agricultural land throughout the world.

Erosion can thus be a direct threat to the farmer. Systems and practices have been developed with a view to conserving the soil, which is to say, to avoid it being moved from one place to another. Evidence of these ideas for conserving the soil against erosion by water, include crops sown along contours and the use of ridges and ditches also constructed along the contours to avoid the water running down the slopes. Tremendous efforts were made to build terraces. It was also recommended not to leave the soil surface uncovered, but to retain residues or some other covering on the surface to absorb the kinetic energy of the wind and the water. In summary, considerable efforts were made to prevent by mechanical means that the forces of wind and water remove the soil.

Water conservation

However, it was not yet recognized at the time that erosion is a consequence of the manner in which agriculture is treating the soils, particularly when the operations are mechanized, rather than a cause.

T. Friedrich
Food and Agriculture Organization of the United Nations (FAO)
Rome, Italy

A case in western Nicaragua can serve as an example. This zone contains some of the most fertile soils of the country and has always been intensively cultivated, during the last 40 years being converted to cotton production. Until today, soil tillage is undertaken exclusively with disc harrows and hence, erosion problems have appeared. To solve the problem, terraces were built, all strictly following the lines of equal contour height. The terraces had irregular shapes and some were so small that tractors could hardly work in them. These terraces were all cultivated with disc implements, in the same way as had been practised over the previous 20 to 30 years. To further worsen the situation, the tractors had to make several turns within the terraces owing to their irregular shapes. As a result, today all the soils in western Nicaragua are both degraded and compacted. But what is even more serious is that the compaction does not allow infiltration of the water (Kayombo and Lal, 1994). This water is being evacuated through drainage channels to avoid it standing in the terraces. The result is not only the formation of enormous gullies that cross the zone, but also a rapid fall in the level of the water table.

This case demonstrates that the loss of soil through erosion is only part of the problem. The loss of water that fails to infiltrate into the agricultural soils can cause even more serious problems over the long term.

It is clear from these examples that drastic changes are needed in the manner of undertaking soil tillage. Runoff, erosion and soil loss cannot be combated through mechanical means, rather they must be tackled through a living and stable soil structure. Only this will allow the rainwater to infiltrate as much as possible into the soil, rather than running off over its surface.

THE CONCEPT OF INTEGRATED MANAGEMENT – CONSERVATION FARMING

Tillage as viewed within the concept of conservation farming

Unfortunately, there is no mechanical implement available that is capable of creating a stable soil structure. Mechanized tillage can only destroy this structure. A new concept of tillage is therefore needed and above all, a profound knowledge concerning the nature of the interventions which can be achieved with each implement.

Naturally, there are differences between the different types of soil as regards their susceptibility to losing their structure. But a stable and optimum structure for plant growth and one that ensures good infiltration, minimizing soil losses due to erosion, can only be achieved through biological processes such as the formation of humus in the soil.

When to cultivate the land

According to the concepts expounded above, the best form of mechanized tillage would be not to do any at all. However, zero tillage does not work well in all cases. Agriculture signifies intervention in the natural processes and hence it must be accepted that in some cases, intervention and corrective action must be taken. Even in zero cultivation, a type of tillage is made in the form of traffic of the machinery over the field to sow, control pests and to harvest. Traffic signifies compaction and this constitutes a type of tillage.

On every occasion when a problem occurs requiring some form of tillage intervention, one must question what is the problem and how best might it be controlled in a way that least affects the soil.

In tillage, five basic operations may be distinguished:

- inversion
- mixing
- fracture
- pulverization
- compaction

There is also a second group of some agricultural operations which have a direct effect on the soil such as:

- mechanical weed control
- surface shaping (ridging, levelling)
- harvesting of underground crops (potato, beet, groundnut, etc.).

Each tillage implement accomplishes specific operations. Knowledge of these and of the availability of suitable equipment can allow limiting interventions to a minimum. Some of the operations of the second group cannot be avoided, but the majority of operations from the first group are unnecessary in agriculture. This is particularly true for the operation of soil inversion, which is precisely the most drastic operation carried out on the soil.

Inversion

This operation inverts the soil in the tillage layer, meaning that it incorporates or buries the surface soil material and brings material from the lower part of the tillage layer to the surface. The need to bring material to the surface from beneath is limited to very special cases. The argument that ploughing controls weeds is not valid when this takes place every year. In this manner, the same number of weed seeds is brought up to the surface every year. The use of the plough was justified in days when tractive effort was limited and when only simple equipment was available for seeding and which needed a clean soil surface.

Mixing

This operation homogenizes and mixes all the soil material down to a predetermined depth. In some circumstances it can be justified, for example to facilitate the incorporation of stubble in temperate climates. The depth of mixing is normally limited to about 10 cm.

Fracture

This operation breaks up compacted soils, opening up cracks and loosening the clods without moving them. In situations where the soil is compacted by machinery or soil with an unstable structure, this operation opens up enough pores in the soil to allow water infiltration. However, the residual effect of fracture varies a great deal and depends on the soil characteristics and on the subsequent treatments (Kayombo and Lal, 1994).

Pulverization

This operation is used mainly to break up the clods and soil aggregates and to form a layer of fine particles approximating to the seed size. It is the operation for seedbed preparation. The operation is carried out in a very superficial layer of the soil. Pulverization at depth is never justifiable, such as is achieved when using a rotary cultivator (rotavator) or a disc harrow. Machinery exists today, which can sow the majority of agricultural crops without any need to pulverize the seedbed. Only in very few cases, for example as in horticulture, is a fine seedbed preparation necessary.

Compaction

This operation is needed after deep tillage and is undertaken shortly after sowing. The soil is compacted to guarantee capillary contact with the underground water. On a reduced scale, compaction is also done during the seeding process after placing the seed in the soil, so ensuring good contact with the soil moisture.

How to treat the land

In the previous sections, evidence has been presented showing that soil does not need any tillage in order to create an ideal structure; indeed, one should limit mechanical interventions in the soil to the minimum possible. However, some agricultural operations cannot be avoided, such as sowing, cultivation operations, fertilizer application, pest control and harvest. Inevitably, these operations lead to soil compaction, and some soils can later be restored easily whereas others cannot. In any case, the machinery operators must be conscious of this situation, organizing for the movement of machinery to be kept to a minimum. Selection of appropriate equipment such as track-laying tractors (Erbach, 1994), soft tyres with low pressure (Vermeulen and Perdok, 1994) and choosing the appropriate moment to enter the field, for example avoiding excessively wet soils, all assist in minimizing negative effects to the soil (Larson *et al.*, 1994).

An interesting way to avoid unnecessary soil compaction is through “controlled traffic”. Ideally, all equipment used by the farmer should work with the same track width setting. In this manner, very compact but narrow strips are established in the field and serve to carry the traffic. In the remaining area, the soil is not compacted, advantages being reflected in the greatly reduced tillage requirements (Taylor, 1994). However, this system often requires a complete change of the machinery used on the farm and strict discipline on the part of the machinery operators.

Two aspects must be taken into account as regards compaction:

- the surface contact pressure: this can be high in the case of trampling by animals but very low in the case of track-laying tractors. This pressure determines the degree of compaction;

- the total compaction load: this can be low in the case of animals but high in the case of tractors, machinery and trucks and it is this that determines the depth of the compaction.

Significance for the farmer

Small farmers

Because the most damaging effects on the soil result from high working speeds and from tillage equipment operated by the tractor power take-off shaft, the problems are less pronounced when working with animal traction. In addition, the effect on the soil is very limited in terms of depth when using draught animals.

This does not imply that working with animal traction would always avoid soil erosion and degradation. The origin of these problems lies not in poor use of the technique nor in the use of the wrong technique, rather it originates because the manner for cultivating the soil is inadequate. For example, if the vegetation is removed from a slope in order to establish a crop, it matters little how it is done because inevitably, erosion will be caused.

Equipment is now available for direct drilling using animal traction within a zero tillage system. However, the equipment is too expensive and sophisticated to justify its purchase by small farmers practising subsistence farming.

Mechanized farmers

For the mechanized farmer, the concept of careful tillage within a conservation farming system implies the need to have access to more specific equipment and in the majority of cases, additional implements will be required. A smaller range of implements, limited basically to seed-drills, fertilizer applicators, some other sophisticated equipment and harvesters would be sufficient only for those farmers in very special circumstances who only grow a limited range of crops. All other farmers are going to need additional tillage equipment because they must always be prepared for additional interventions which might be needed during difficult times or on other occasions (Gogerty, 1995).

A farmer who has access, for example, to a disc plough or a disc harrow, might also need a subsoiler, a chisel plough, a moldboard plough and other implements, depending on the type of soil and the climate (Reynolds, 1995). However, much of this equipment will not even need to be used every year. This implies that the farmer, at first glance, will have a very high level of investment tied up in his machinery.

In addition, other changes will be needed in the farmer's machinery pool. Even under a system of reduced tillage or zero tillage, additional quantities of crop residues will exist on the soil surface. The technology for drilling therefore should be adapted to these new circumstances, which implies the purchase of seed-drills suitable for the different crops. For row crops, there will also be a need for inter-row cultivators that allow mechanical weed control to be undertaken whilst still leaving the residues on the surface.

These changes are important and above all, expensive and risky for the farmer. Without specific technical assistance and other incentives, it will be difficult to generate a process for such changes.

TECHNICAL TILLAGE PARAMETERS

Speed effects

Performance

Apart from the working width, forward speed is the factor that allows increased rates of operational performance. Whilst for animal traction systems the speed is more or less limited by the type of animal used, in the case of tractors a very wide range of speeds is possible. On many occasions the farmers and particularly the machinery operators, cannot resist increasing forward speed as this is the easiest and cheapest way in which to increase performance. With this approach, they often exceed the recommended speed limits for each particular operation.

Action on the soil

Each implement is designed to perform best within a particular range of speeds. For moldboard ploughs with cylindrical and vertical bodies, the recommended speed range is 4 to 5 km/h. For helicoidal and inclined bodies, this can reach 10 km/h but at higher speeds the soil will be excessively pulverized and thrown too far. In contrast, the chisel plough and the spike-toothed harrow only work well if the forward speed is in the range of 8 to 12 km/h because they break up and mix the soil clods through impact. On the other hand, toothed harrows or disc harrows used with draught animals do not have much effect on soil disintegration, only on land levelling.

Energy consumption

With increased working speed, the draught force increases quadratically and hence, also the energy requirement. This is reflected in the tractor fuel consumption rate and hence, in the level of operational costs. For this reason, increasing speed is not an appropriate method for increasing tillage performance rates. Doubling forward speed (for instance to 8 km/h rather than 4 km/h), performance doubles but four times more energy and fuel will be required.

Conclusion: whereas in the case of using animal traction the working speed is limited, which does not allow some implements to accomplish their complete action, in the case of the tractor there can be a problem of excessive speed with negative repercussions on the soil structure and energy consumption.

Depth of work

Tillage may be differentiated according to the depth of work. Each type of implement has particular characteristics and defined needs.

Subsoiler

The subsoiler reaches down below the ploughing depth to break up a compacted layer that is beyond the reach of normal tillage equipment. The operation serves to form cracks that will improve infiltration rates and root penetration. The working depth of the subsoiler should be decided according to the compaction that has been identified and the soil moisture content at this depth.

The operation of subsoiling requires a great deal of energy. For this reason, it is not suitable for animal traction. With a tractor one must consider it as an expensive soil improvement operation, and one which should not be carried out routinely.

Great care must be taken, particularly in unstable soils, not to recompact the soil immediately after subsoiling as this can cause compaction at depth that is even worse than that existing before the operation. There are also some silt soils where there is a danger that the fine material might accumulate in the cracks and form further compaction through sedimentation. In general, before carrying out subsoiling one must determine the origin of the compaction and plan a way to change those conditions, for example by stabilizing the new loosened structure with a deep rooting crop.

Primary tillage

Primary tillage is the traditional tillage operation that modifies the upper soil layer (the Aphorizon) over the entire cultivated surface. This serves to eliminate surface compaction, open up the soil and create a clod type structure to accumulate water and often also, to incorporate weeds and weed seeds by ploughing.

The depth of primary tillage depends on the tractive force available. Using power from animal traction, the depth is normally between 10 and 20 cm whereas with tractors, particularly in view of the increased power of modern tractors, ploughing is done down to a depth of 40 cm in some countries.

There is much controversy concerning optimum ploughing depth. Generally, one should not increase the working depth just because sufficient power is available. For soils with a thin layer of topsoil, deep ploughing can literally destroy the soil, something that frequently occurs with the use of tractors. The increase in harvested yield that occasionally occurs after deeper tillage, proves to be unsustainable in most cases. This depends greatly on the fertility and depth of the soil. On the other hand, with good soil structure the plant roots reach the deeper parts of the soil without any need for deep tillage. Over the long term, deep tillage consumes more fuel whilst benefits are not assured.

Secondary tillage

Secondary tillage serves to prepare the soil for sowing. This includes shaping the surface, levelling, making ridges or furrows for irrigation and establishing a seedbed. This seedbed should extend only over a shallow horizon down to the depth selected to receive the seed. Normally, secondary tillage operations level and pulverize the soil and working to greater depths will only bring about an unnecessary loss of soil moisture. In cases where the soil of the seedbed is loose, one must also include within the secondary tillage plan an operation to compact it again.

The need to make a traditional seedbed arises from the inadequacy of the traditional techniques for sowing into virgin soil. But at present, technology is sufficiently advanced to allow sowing in the majority of soils without any prior tillage. In addition, one should realize that secondary tillage in a tropical climate leaves a pulverized surface in critical condition, with a serious risk of erosion.

Cultivation, weeding

This type of tillage is normally superficial and serves to control weeds, incorporate fertilizer, break up surface crusts and to ridge. The functions of this type of tillage and the correct equipment selection will depend upon the particular problem and the kinds of weeds encountered. To remove the weeds, the basic functions are to pull them out and leave them on the surface, or to bury them and cut the roots. Great care must be taking when adjusting the equipment to avoid damaging the crop. A common fault for instance, is to let the weeds grow too much and then to try to control them with deep cultivation. In the case of maize, this destroys all the surface roots of the crop, that are the most important ones for its nutrition.

Soil characteristics – texture and moisture

The soil characteristics have a great influence on the selection of the type of implement, its rate of wear, the power requirements for the tillage operation and the time available for the cultivation operation.

Moisture content and the tillage window

For each soil type there is a specific optimum moisture content for tillage. However, the optimum ranges of soil moisture content for tillage, or the "tillage window", may be either pronounced and narrow as in the case for clay soils, or less pronounced and broad as for sandy soils. In general, one should look for a "tillage window" to achieve optimum results with acceptable energy costs. In some soils, this "window" can be extremely narrow and it can practically preclude the possibility of tillage. Direct seeding and zero tillage techniques are more appropriate in these cases. Whilst the use of draught animal power is necessarily limited to tillage within the optimum "tillage window", the tractor allows one to work either side of these limits under extreme circumstances. However, this may result in damage both to the soil and to the equipment.

Dry and moist tillage

Departing from the optimum moisture range for tillage can be justified in certain cases. From the soil point of view, it will have the following effects:

- In sandy soils, tillage undertaken in dry soil will not have the desired effect because the sand has no cohesive forces. For example, a plough will not invert the soil; it will only form a furrow. When the sandy soil is too moist, there is less danger of compaction than in heavier soils but it can still lead to serious problems.

- Silty soils can be tilled when dry. However this consumes more energy than tillage under moist conditions. In addition, it can cause dust and consequent soil loss through wind erosion. One should avoid all tillage of silty soils under conditions that are too moist, as serious compaction will result.

- Clay soils are practically impossible to till when dry. It requires too much tractor power and causes damage to both the implement and the tractor. In addition, dry tillage produces huge clods that are extremely difficult to break down later. Tillage under conditions of excessive moisture causes wheel slippage and serious compaction problems.

- Tillage of heavy soils when dry may be justifiable as a method for soil improvement, particularly if this concerns subsoiling. The effect of subsoiling is more pronounced in dry soils and the rupture zone is larger. Afterwards, the soil is left exposed to the elements so that the atmospheric forces cause the break-up of the clods due to changes in their temperature and moisture content.
- Tillage in moist soil combined with puddling. The soil when wet is a special case applied to irrigated rice production.

Abrasion

The abrasion of the implements and the consequent implement wear rate depends on the texture and the geological origin of the soil. In general, light and sandy soils are more abrasive than heavy clay soils.

Chapter 5

Tillage implements

MOLDBOARD PLOUGH

Mode of action, forces and adjustment

The moldboard plough is one of the most classic soil tillage implements after the wooden plough. Whilst the wooden plough works as a chisel, the moldboard plough was developed so as to cut a section of soil and to invert it through approximately 130°. The moldboard plough is the most appropriate implement to invert the soil but its action of mixing the soil is very limited.

The forces acting on the moldboard plough can be subdivided into three components: the longitudinal component of soil resistance, the lateral component given by the lateral acceleration of the soil section and the vertical component determined by the plough shape and acting downwards. These forces are compensated through the line of pull, the operator (or the tractor) and parts of the plough itself such as the tail of the heel and the landside which support part of the lateral and vertical forces. Adjustment of the plough is done in such a manner that the lateral forces are neutralized by those on the plough components and the line of pull. The vertical forces can partially be increased by the operator of the animal traction plough or by the tractor.

This signifies in practice that if the operator of an animal traction plough has to push the plough to one side or apply any force, the plough is incorrectly adjusted. The same is true for the tractor: the plough should follow the tractor in a straight line without any need to adjust the chains on the lower hitch links or to correct the direction of advance with the steering wheel.

There are two main groups of ploughs for tractors: mounted or semi-mounted ploughs and trailed ploughs. The mounted and semi-mounted ploughs have the advantage of transferring part or all of the plough weight and the vertical force to the tractor, so improving the traction. The automatic adjustment provided through the hydraulic system allows maintaining constant the working depth or the draught force of the implement. Unfortunately, there are very few operators who know how to correctly use the hydraulic control system.

T. Friedrich
Food and Agriculture Organization of the United Nations (FAO)
Rome, Italy

Types of moldboard plough

- Ploughs for animal traction: there are two common types of animal traction plough depending on whether or not they have stabilizers. Those without stabilizers have no wheel or skid whereas as those with longitudinal stability are equipped with simple support wheels or skids. The heavier ploughs with both lateral and longitudinal stability, which might consist of an axle with two wheels, are less commonly used. As regards the single support wheel, a simple skid is often preferable, particularly in clay soil and where there is little abrasion. The wheels that are most commonly fitted are very small and hence do not run any better than a skid, but they are more expensive.

- There is a wide range of different types of tractor plough including large deep-penetrating models, lightweight ploughs, simple and reversible models. It is important to select a plough with a furrow width greater than the tyre width of the tractor. The width of cut of the furrow determines also the maximum depth that can be achieved with a particular plough. The working depth cannot be more than between 0.8 and 1 times the working width. For this reason, shallow ploughs have many small bodies whereas deep ploughs will have a few wide bodies.

- There is a wide range of types of moldboard according to soil type, the intended use and the working speed. There are also slit moldboard bodies and those coated with special steels, teflon or other synthetic materials designed to reduce resistance in sticky soils.

- A special form of moldboard plough is the ridging plough to shape ridges. In some cultivation systems with animal traction and when sowing along furrows or ridges, the ridging plough is the only tillage implement used on the farm.

DISC IMPLEMENTS

Mode of action, forces and adjustments

In this section all disc implements will be described of which all have basically similar operating principles. The disc, depending on the angle of attack, also cuts a section of soil and inverts it. However, because of the movement of the disc, the acceleration differs according to the position of the disc and the resultant internal friction. The soil is thus also pulverized and mixed.

Whilst the disc does not invert the soil as well as the moldboard body, it both pulverizes and mixes the soil at the same time. In addition, disc implements tend to be less susceptible to damage from stones and stumps and therefore are well adapted for less cultivated land. For these reasons, being very universal and robust, disked implements have been very successful in mechanized tropical agriculture. However, within the concept of conservation farming and more careful and managed tillage, disked implements should be considered very critically.

The forces acting on the disc can also be subdivided into three components. The longitudinal component has very approximately the same value as for a moldboard plough; the lateral component can be very large; and the vertical component acts upwards, which is opposite to a moldboard plough. These characteristics also have consequences:

- In order to support the lateral forces, disc ploughs need a very strong support wheel, whereas disc harrows are designed with two sets of discs acting in opposite directions.

- The disc only penetrates the soil due to its weight as the vertical force is acting upwards. In the case of heavy soils, one has to increase the overall weight of the implement by adding additional ballast weights. For these reasons, disc ploughs tend to be made heavier and are not well suited for use with work animals.

These characteristics of the disc are the reasons for the problems of soil degradation that can often be observed in zones where disc ploughs have been misused. The pulverizing action of the disc brings about a loss of soil structure, more rapid mineralization, increased erosion and loss of moisture and poor infiltration of the water. The disc enters into the soil due to its weight until such depth that the vertical soil force is equal to the implement weight. This means that the disc supports itself on the soil and can be considered as a roller compactor of the subsoil. In zones where disc harrows are frequently used, one can find very compacted horizons beneath the normal working depth of the implement. These compacted layers inhibit the infiltration of water and thus quickly cause drought conditions to build up and can also contribute to a process of desertification on large areas over the longer term.

As regards adjustment, one must distinguish between the two types of disc implement: with individual discs such as disc ploughs, or with the discs mounted on a common shaft such as disc harrows.

In the first case, that of the plough, both the vertical and horizontal angles of the disc may be adjusted. These adjustments allow adapting to the type of soil and also adjustment of the degree of soil pulverization and the ease of penetration into the soil. In the case of the disc plough, correct adjustment is achieved in just the same way as for the moldboard plough. This means that when correctly adjusted, all the lateral forces on the plough are balanced and the plough proceeds in a straight line without need to adjust the chains on the lower links of the three-point linkage system.

In the case of the harrow, one can only adjust the horizontal angle. With this, and by adding additional ballast weights, one may adjust the depth of work and the degree of soil pulverization.

Types of disc implement

- Because of the lateral forces required, there are very few disc ploughs for use with draught animals. The only exception is the disc harrow, which exists in a few countries.

- Disc implements are probably the most commonly used types of tillage implement used in tropical countries. This group of implements can be sub-divided as follows:

  - ❑ disc ploughs exist in various versions as either tractor-mounted or trailed, simple or reversible. Reversible disc ploughs, because of the symmetry of the disc, are much simpler than their counterpart, the reversible moldboard plough. With the reversible disc plough, all that is required is simply to turn the plough body through a vertical axis when changing direction;

- an intermediate form between ploughs and harrows is known variously as the one-way disc plough, the rangeland plough or the plough-harrow, which has the discs mounted on a shaft and which work only towards one side as in a disc plough. There are mounted and trailed models, in simple form or combined with seed-drills for direct drilling;
- disc harrows are always made up of gangs of discs of equal number that work in opposite senses so as to neutralize the lateral forces. There are both mounted and trailed models, large ones for tilling virgin land or small lightweight models for secondary tillage. Disc harrows are very popular, robust and versatile but they also contribute most towards the degeneration of agricultural soils on a world-wide scale;
- disc implements are today becoming very popular for ridging up the soil, for making furrows and for tied ridges.

- Discs with a slight curvature or flat are used for seed-drills for zero tillage and with other implements to cut through the soil and the residues so as to deposit the seed or fertilizer.

CHISELS

Mode of action, forces and adjustments

Chisel ploughs, due to their mode of action, are the tillage tools that most closely simulate the wooden plough. When the chisel is first introduced into the soil, compression is built up. The soil finally bursts upwards leaving a fracture zone fanning outwards from the point of the chisel at an angle of approximately 45° in dry soil. The chisel thus serves to break up the soil mass. Chisels used with animal powered systems are limited in practice to this type of action.

When higher forward speeds are employed, the soil is also moved to the sides. Selecting certain shapes of chisel point can assist this action. For this reason, tractor operated chisel ploughs used at speeds of between approximately 10 and 12 km/h have a good mixing action. The impact of the chisel on the aggregates and clods also brings about a pulverizing action on the soil. However, the effect is not very pronounced in loose soils. It follows that repeating passes with a chisel plough in loosened soils does not cause major soil pulverization.

The chisels leave undulations both on the surface and at the bottom of the disturbed layer, where rupture starts from the share point, spreading out at an angle of 45°. It is therefore recommended, when using chisel ploughs, that at least two crossed passes should be made so as to level the soil surface.

The forces acting on the chisel greatly depend on the shape and particularly, on the angle of attack. A shallow (small) angle of attack improves penetration and reduces the draught force. In addition, it improves the bursting and the mixing of the soil because part of the soil material is lifted from the lower horizons towards the upper ones.

This characteristic can be a disadvantage in certain situations where the chisel point touches moist material in the lower layers and carries it to the surface in the form of small cylinders or clods which will be difficult to break down later.

Whilst the simple chisel does not require much draught force and can be used by work animals, the use of gangs of chisels for homogenizing and mixing the soil at high speed requires powerful tractors. This results in the need for the chisel implement to have a working width at least equal to the track width of the tractor and for it to be used at a high speed.

Vibratory chisels mounted on springs serve to generally improve the pulverizing action and to pull out the weeds. Generally, these are used for secondary tillage operations working to a depth of about 15 cm, whilst rigid chisels are used for primary tillage and for subsoiling.

Types of chisel

Rigid chisels

- Subsoilers are large and robust chisels, which can reach down to a working depth of more than a metre. Because of the tractive force required, their use is restricted to tractors, not animals. One may distinguish between the traditional (vertical) shape, the parabolic shape and the "Paraplow". The vertical shape only works well in dry conditions and requires more tractive effort than the parabolic shape. However, the parabolic shape has the disadvantage of bringing up clods to the surface. To avoid this inconvenience, inclined parabolic tines are now used. The "Paraplow" has similar characteristics. This only needs a relatively small draught force when compared to other subsoilers, it leaves a subsoil profile that is more level than that left by a normal chisel plough and its action is totally limited to bursting or breaking up the soil. In order to improve the bursting action and levelling of the working surface that is achieved with vertical or parabolic tines, one may use tines with spread wings.

- Chisel ploughs for primary tillage with animal traction are available, with a maximum of three chisel tines. Others are available for use with tractors. They are used to break up the plough layer and if tractor power is used, to mix it. Depending upon the type of soil and the desired effect, they can be equipped with a variety of different points. However, narrow points and tines are normally used for primary tillage.

- There are also cultivators with rigid tines for secondary tillage and weeding operations, designed either for animal traction with five narrow tines or broader duck's foot tines, or designed for tractors. The points in the case of tractor-operated implements may also be shaped as duck's feet or as spread wings, with widths of up to one metre for the control of superficial weeds in arid zones.

Vibratory chisels

This type of chisel is used both with animal and tractor powered systems. The heavier versions are used to mix the soil, the lighter ones for secondary tillage, seedbed preparation and weeding.

SPIKE-TOOTHED TINES – LEVELLERS AND HARROWS

Spiked teeth are types of chisel that are used for levellers and harrows. Their mode of action is very similar to that of the chisel, with the limitation that the teeth are always vertical. For this reason, the depth of work of spike-toothed levellers depends on the weight of the implement and the angle of pull.

In the same way as with the chisels, spike-toothed levellers for animal traction are limited to levelling the soil surface, whereas tractor operated models, due to their higher working speed, can break up and to a certain degree, pulverize the soil aggregates through impact.

Again as in the case of chisels, this soil pulverization effect cannot be increased by repeated crosses over the same terrain. On the contrary, repeated passes of the leveller will bring about an effect of classification of the aggregates through which the finer particles will descend through the profile and the larger aggregates will rise to the surface where they cannot be further broken down.

ROTARY CULTIVATORS (ROTAVATORS)

Rotary cultivators (rotavators) are examples of implements powered through the tractor power take-off shaft. Normally, rotary cultivators rotate in the direction of advance of the machine. This means that little tractive force is needed by the tractor but only energy to operate the implement and they can thus be used with very lightweight tractors.

These rotavators are very popular for work in horticulture although they are also used in agriculture, above all for pulverization of heavy soils. With this objective in mind, they are often combined together with a seed-drill in machines for land preparation and sowing in a single pass as used in direct seeding operations.

These implements should be considered very critically under the climatic conditions of the tropics due to their strong impact on soil structure and the consequent high erosion risk.

ROLLERS

Rollers are important implements but in some tropical countries they are virtually unknown. The types of roller are quite different according to their use:

- smooth rollers compact the surface and are used to recompact pasture or roads, or to prepare the soil surface for crops such as alfalfa for seed production or mechanized lentil production;

- "cross-kill" type rollers with surface profile or roller-packers consisting of a set of rings, are used to break down the clods. They are available either as separate implements as rollers or in combination with chisel ploughs or tined cultivators and are designed to pulverize and to recompact the surface of a seedbed. For heavy soils in particular, it is recommended to pass a cross-kill roller immediately after the plough to break down moist clods before they become dry and hard.

- Packers or compactors for the subsoil are a special form of clod-breaking roller. They consist of a series of annular rings that rest on their edges in the soil and so compact it from the bottom upwards. They simulate the natural recompaction process of the soil and are normally used in combination with a plough or a seed-drill.

DIRECT DRILLING - ZERO TILLAGE

Concepts

The concepts of direct drilling and zero tillage both represent a type of soil conservation tillage or minimum tillage. Whilst “zero tillage” clearly excludes any type of tillage, the term “direct drilling” may be interpreted in various ways.

One way would be to combine all the conventional tillage operations into a single operation that includes sowing. Interpreted in this way, direct drilling brings advantages in terms of reduced traffic and compaction, simpler work organization and perhaps in the costs of land preparation, the time necessary to sow and the reduced exposure of the soil to the elements. However, it often includes intensive tillage with implements operated by the power take-off shaft and needs large and powerful tractors.

In contrast, zero tillage does not include any type of tillage and can be done manually, by animal traction or by a small or large tractor. In this system, the seed is placed directly into the soil by injection or with direct drills with disc or tined furrow openers which cut through the stubble, open up the soil and deposit the seed. The technology of zero tillage available nowadays allows the use of this concept for practically any type of crop.

Equipment

Direct drilling

Two basic types of equipment may be used for direct drilling:

- a special combination of the same implements used traditionally for this task. For example, a chisel plough mounted at the front of the tractor and a combination of rear mounted implements for seedbed preparation and seeding. Alternatively, a short tined chisel plough, a rotary harrow and a seed-drill, all rear-mounted behind the tractor. In this case, all the implements are simple but are hitched together for this particular task.

- special purpose implements for direct drilling. These generally consist of chisels for breaking up the soil at depth, a rotary cultivator and a seed-drill. These equipment components form a single machine.

Zero tillage

For zero tillage, implements are used which place the seed in the soil without any type of tillage operation being involved:

- when being done by hand, equipment may range from a simple stick to punch holes for placing the seed to a hand-operated seeder-planter to inject the seed and sometimes, also the fertilizer;
- zero tillage seed planters are available for use with work animals and plant one or two lines for row crops. They work with openers which consist of discs or star wheels;
- tractor operated seed-drills are available for reseeding pasture, sowing cereals and for row crops. Depending upon soil conditions, they work with chisels, simple discs or double discs and also with star wheels. The use of double disc openers is the most common arrangement. The equipment is generally very heavy so as to ensure uniform penetration to the desired seeding depth in hard soils. The distance between the openers in a single gang should not be too narrow so as to ensure that the stubble can pass between them. For this reason, the openers are placed in two or three lines or gangs in order to achieve a minimum inter-row spacing of about 15 cm.

Chapter 6

Implements and methods for the preparation of agricultural soil

Dryland agriculture as practised in the semi-arid region of Northeast Brazil, is mainly undertaken using a hand hoe for preparing the soils and to open up holes to receive the seed when the soil is sufficiently moist. In this case, there is no conventional soil preparation (ploughing). The soil is only disturbed superficially by hoeing to eliminate the weeds and to reduce moisture loss due to evaporation.

Another practice is to broadcast bean seeds (*Phaseolus vulgaris L.*) in the regions where rainfall is strongly influenced by the Atlantic Ocean and characterized by regular rainfall. The crop seeds are hand broadcast along alleys that have been opened up previously in the native vegetation, grasses and bushes and which, after the seeding operation, will be cut down to give room for the crop. This constitutes a cultivation system without any previous soil preparation, where the native vegetation springs back to recover alongside the crop. After the harvest, the land will stay fallow for two or three years. This system is ecologically sound and forms part of the programme of the Government of Pernambuco State. It is called “Mata Viva” (living shrub) and has the objective of protecting the soil from erosion in the mountainous regions receiving abundant rainfall, thus conserving the soils and the natural resources of the environment.

In the Northern and North-eastern part of Brazil, the farming system, particularly the soil preparation for seeding, differs from that practised in other parts of the country due to the rains being more intensive. In the temperate regions with moderate rainfall, the crop is planted in the bottom of the furrow where there is more moisture available in the soil. This would also be ideal for the tropical region but because of the rainfall characteristics showing high intensity over short periods, this is not possible. The solution is to sow on the ridges (Kepner*et al.*, 1972).

Despite the well-known techniques of conservation agriculture to protect the soil, today in Brazil shifting agriculture is still practised. It is a primitive method to cultivate the soil which consists in cutting down the forest and then burning it to facilitate planting the crops. The area is then abandoned once it becomes unproductive and from there, the farmer leaves to look for new areas which have not yet been exploited. This type of “slash and burn” agriculture is common in the Northern (Amazonian) region and is known as migratory agriculture (Kitamura, 1982).

J. Barbosa dos Anjos
Brazilian Enterprise for Agricultural and Livestock Research (EMBRAPA) -
Centre for Agricultural and Livestock Research in the Semi-arid Tropics (CPATSA)
Petrolina, Brazil

OBJECTIVES OF SOIL PREPARATION

The objectives of soil preparation are based on the following principles (Mazuchowski and Derpsch, 1984):

- elimination of undesirable plants, reducing competition with the established crop;
- achievement of favourable conditions for sowing or for placing vegetative material into the soil, so allowing germination, emergence and good plant development;
- maintenance over the long term of fertility and productivity, preserving the soil organic matter and avoiding erosion;
- elimination of hard pans or compacted layers to increase water infiltration through the soil whilst avoiding erosion;
- incorporation and mixing of lime, fertilizers or agro-chemical products into the soil;
- incorporation of organic and agricultural residues;
- land levelling to facilitate better quality of work with machinery during sowing and up to the time of harvest.

The option chosen concerning the type of soil preparation depends on many factors and for each situation, particular decisions must be made at farm level. Each operation also involves particular time commitments depending upon the power source to be used, be this manual labour or animal traction (Table 10).

TABLE 10
Work rates per unit area needed to carry out a selection of agricultural tasks on the farm

Implements and tools used on the farm (animal traction and hand-operated)	Time required (hours/hectare)
Plough (animal traction)	20
Tined harrow (animal traction)	6
Disc harrow (animal traction)	4
One-row planter for beans (animal traction)	10
Bean planter (hand planter) “matraca”	16
One-row maize planter (animal traction)	6
Maize planter (hand planter) “matraca”	8
One-row rice planter (animal traction)	11
Rice seeder (hand seeder) “matraca”	18
“Planet” type seed-drill (animal traction)	8
“Planet” type cultivator (animal traction) + hand hoe	40
Hand hoe (for weeding)	80
Manual harvest of maize (40 sacks)	60
Manual harvest of beans (15 sacks)	80
Manual harvest of rice, reaping with a small hand sickle (35 sacks)	64
Manual threshing of beans with a wooden flail (15 sacks)	30
Manual threshing of rice with a wooden threshing table (35 sacks)	35

- When calculating the number of hours needed to prepare a hectare of land with mechanical power or animal traction, the parameters of working width, forward speed and total width of the working area must be considered. Equation (1) may be used to calculate the time needed for manoeuvres (Tm).

$$Tm = \frac{\frac{L}{l} \times t}{3\,600} \times f \qquad \text{(Equation 1)}$$

where:

Tm = time lost in manoeuvres (h/ha);
L = width of the area (m);
t = time needed for one turning (seconds)
l = working width of the implement (m)
f = factor $\left(\frac{100}{L}\right)$

- The effective time (h/ha) is calculated on the basis of Equation (2)

$$Te = \frac{10}{l \times V} \qquad \text{(Equation 2)}$$

where:

Te = effective time (h/ha);
l = working width of the implement (m);
V = forward speed of the implement (km/h).

- The operational time (h/ha) is the sum of (Tm + Te) according to Equation (3).

$$To = Tm + Te \qquad \text{(Equation 3)}$$

where:

To = operational time (h/ha);
Tm = time lost in manoeuvres (h/ha);
Te = effective working time (h/ha).

IMPLEMENTS FOR SOIL PREPARATION

- Animal drawn ploughs (mouldboards) are the most commonly used implements, together with tractor operated mouldboard or disc ploughs, although the working efficiency depends more on selection of the method of soil preparation than on selection of the type of implement (Figure 7). Ploughing in strips or bands is a method by which no more than 50 percent of the total land area is tilled. The unploughed part between two strips is used for rainwater collection and for redirecting the water towards the area being cropped

FIGURE 7
Method of ploughing in strips with animal traction

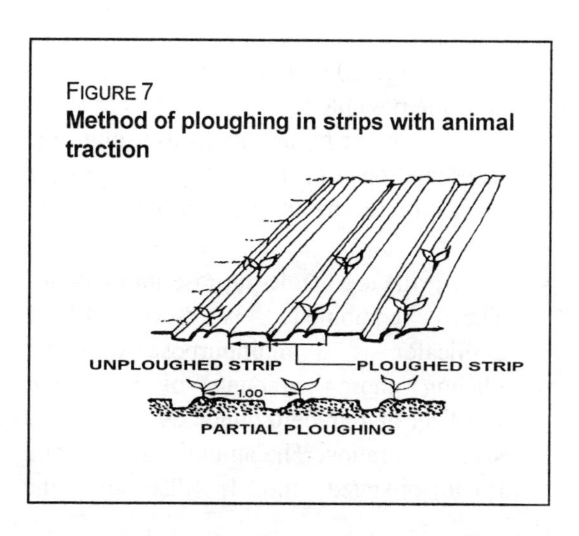

(Anjos *et al.*, 1988). Figure 8 shows the type of lifting share used, which is the same as that employed for harvesting groundnuts. Another method is to use a tractor-mounted reversible disc plough for strip ploughing the land for watermelon production under dryland conditions.

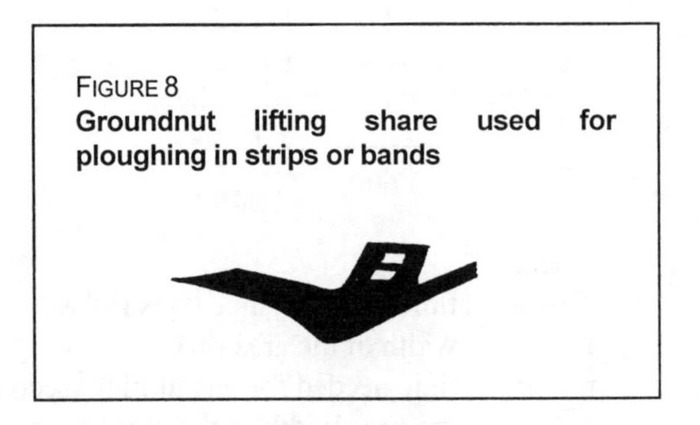

FIGURE 8
Groundnut lifting share used for ploughing in strips or bands

- Harrows are implements used to complete the work accomplished with the plough, breaking down the clods and levelling the soil after ploughing. They can be drawn by draught animals or operated with tractors. Some heavier models are often used as a soil preparation implement to replace the plough and operated by a tractor, but their continuous use tends to degrade the soil causing a compacted plough pan. The harrow components which disturb the soil are the tines (either rigid or flexible) or the discs (straight or serrated).

- Cultivators are used to scarify the soil surface with the objective of controlling weeds and improving the soil physical conditions (Figure 9). This hoeing operation may be considered as a type of minimum tillage when it is undertaken as a cultural operation either before sowing or after plant emergence.

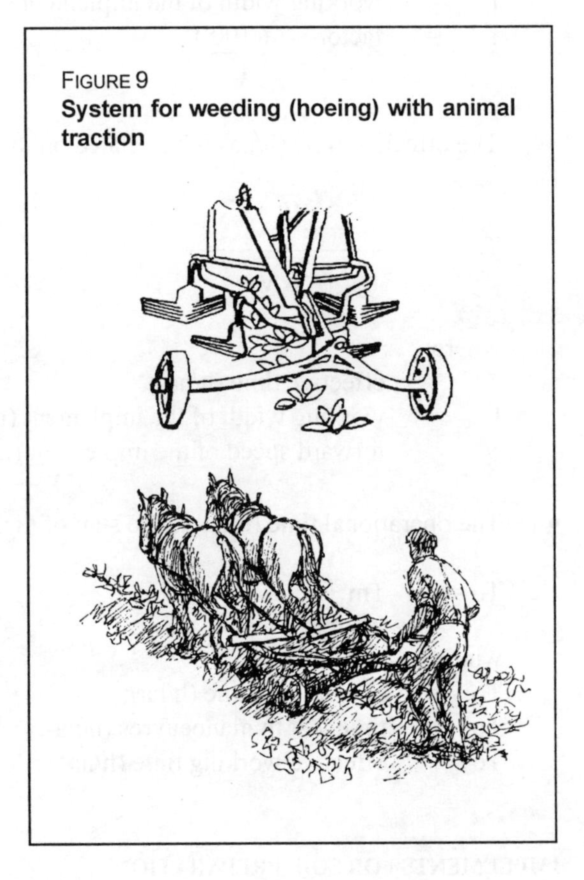

FIGURE 9
System for weeding (hoeing) with animal traction

- Ridgers are designed to open up furrows in the soil (Figure 10), either to serve as water channels, to orient or to mark out the crop rows for fertilizer distribution, or as a tillage operation to control weeds. It is quite possible to couple ridgers to seed-drills in order to simultaneously plant the seed and open up furrows for subsequent irrigation (Franz and Alonço, 1986).

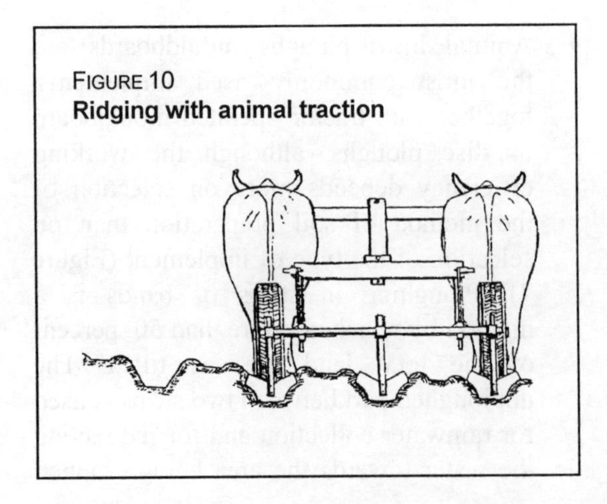

FIGURE 10
Ridging with animal traction

- Seed-drills are single-purpose implements. The combined seed-drill/fertilizer applicator is a multipurpose tool for placing chosen amounts of seed and fertilizer at a predetermined depth and in a single operation. The animal powered and tractor-operated models distribute the seeds and fertilizers in lines to make up the rows. The fertilizer applicator is located so as to apply the fertilizer parallel to and beneath the seed so that it does not affect the seed germination.

- Manual planters are tools used for planting into holes punched in the soil. They can be single-purpose for the distribution of seeds only or multipurpose for the placing of both seeds and fertilizers in a single operation (Figure 11). The fertilizer is placed in the hole to one side and lower than the seed itself. This avoids damaging the seed during the germination process. There are some specific models designed for particular crops, such as the planter for non-delinted cotton seed, planters for groundnut seed (which is very sensitive to mechanical damage), together with special accessories for sowing small graminaceous seed such as buffalo grass (*Cenchrus ciliaris L.*) (Anjos *et al.*, 1983) or for direct planting through stubble or other vegetative cover material (Almeida, 1993).

FIGURE 11
Manual planter for non-delinted cotton seed

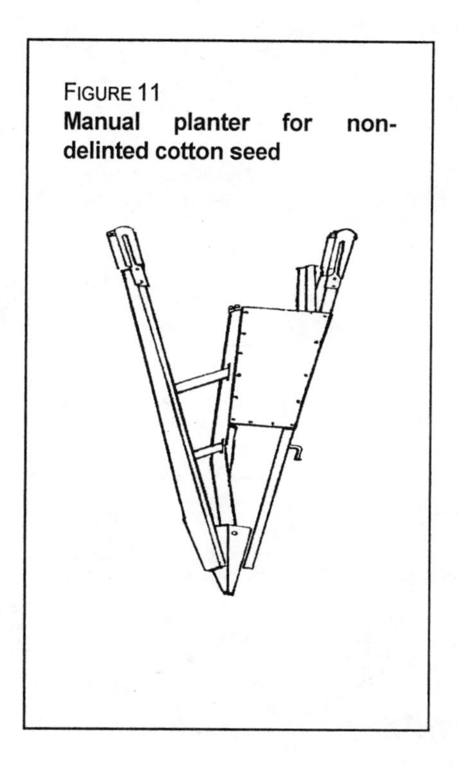

Chapter 7

Effect of tillage on soil physical properties

Cultivation systems, defined as tillage and management systems for the crops and their residues, have an important influence on the physical properties of the soil. To a major extent, the nature and the extent of this influence depend on the soil tillage.

Soil tillage is carried out with the objective of changing the soil physical properties and to enable the plants to show their full potential. Soil tillage techniques are used in order to provide a good seedbed and root development, to control weeds, to manage crop residues, to reduce erosion and to level the surface for planting, irrigation, drainage, cultural tasks, incorporation of fertilizers or pesticides and harvest operations. Incorrect soil tillage due to a failure to understand the objectives and the limitations of tillage techniques, can give rise to negative effects. Incorrect tillage is one of the causes of erosion and of physical degradation of the soil.

Physical degradation of the soil may be defined as the loss of the soil's structural quality. This structural degradation may be observed both on the surface where thin crusts may be seen and also below the surface in or below the ploughed horizon where compacted layers may be formed. With this type of degradation, the infiltration rates of the water into the soil are reduced, whilst the rates of runoff and erosion increase (Cabeda, 1984).

CAUSES OF PHYSICAL SOIL DEGRADATION

The main causes for degradation of the soil physical characteristics are as follows (Cabeda, 1984):

- Inadequate cover of the soil surface. This exposes the aggregates at the soil surface to the action of the rain which brings about structural collapse of these aggregates, forms crusts with a thickness of about 1 mm on average and drastically reduces the water infiltration rate.

- Excessive tillage or tillage undertaken when the soil moisture content is not appropriate causes adverse effects. Excessive tillage on the surface breaks up the aggregates, favouring the formation of surface crusts, increased runoff and the erosive transport of the soil particles. The reduction of the surface roughness leads to less infiltration and increased runoff, increasing the erosive effects due to the increased kinetic energy of the water running over the surface. Subsurface compacted layers, normally situated at depths of 10 to

E. Giasson
Soils Department, the Federal University of Rio Grande do Sul
Porto Alegre, Brazil

30 cm and with a thickness of 10 to 15 cm, are also formed when poorly selected and excessively heavy equipment passes over the soil and when the soil consistency is plastic. These layers offer strong resistance to root penetration and restrict the capacity for water infiltration and aeration.

- Inadequate management causes loss of soil organic matter. This leads to greater soil density and lower moisture retention capacity and stability of the soil aggregates, which together contribute to an overall loss of the quality and stability of the soil structure.

STAGES IN THE PHYSICAL DEGRADATION OF SOIL

The degradation of agricultural soils takes place in three stages (Mielniczuk and Schneider, 1984):

1st Stage The original soil characteristics are slowly destroyed. Degradation is hardly noticeable due to the low intensity of the processes and because productivity is maintained through the use of corrective measures and fertilizers.

2nd Stage More important losses of soil organic matter occur with heavy soil damage (structural collapse). Apart from surface crusting, there is also subsurface compaction that impedes water infiltration and root penetration. In this way, erosion becomes more accentuated and the crops respond much less efficiently to corrective measures and the application of fertilizers.

3rd Stage The soil becomes extensively damaged, with major collapse of the pore space. Erosion is accelerated and there is difficulty to operate agricultural machinery. Productivity falls to minimum levels.

The time to reach this third stage of degradation depends on the intensity of the use of inappropriate tillage and management practices, on the land slope, the soil texture and on the resistance of the soil to erosion by water (Mielniczuk and Schneider, 1984).

PRINCIPAL PHYSICAL CHARACTERISTICS AFFECTED BY TILLAGE

The loss of the physical quality of a soil can be evaluated through the changes that occur to some of its most important physical characteristics such as density, porosity, pore size, distribution, structure and infiltration rate.

Soil density and porosity

Soils naturally possess different densities owing to variations in their texture, porosity and organic matter content. Brady (1974) quotes sandy soils as having a density of 1.2 to 1.8 Mg/m^3 and a porosity of 35 to 50 percent, whilst clay soils have a density in the range of 1.0 to 1.6 Mg/m^3 and a porosity of 40 a 60 percent. However, the density and the porosity vary according to the type and the intensity of tillage, thus they are good indicators of the adequacy of various tillage systems, indicating the serious or minor compaction which the tillage operations may have caused.

Adequate values for soil density were defined by Archer and Smith (1972) as those which provide maximum moisture availability and at least 10 percent air in the pore spaces at a suction of 50 mb. According to the authors, these optimum soil densities are around 1.75 Mg/m^3 for loam soils, 1.5 for sandy loams, 1.40 for silty loams and 1.2 for clay loams.

Modifications to the soil physical properties due to tillage systems can give rise to an increase in soil density, greater resistance to root penetration and a reduction in soil porosity, characterized by a compacted layer below the cultivated horizon. This compacted layer affects the movement of the water and the development of the root system due to the mechanical impediment and the lack of air.

The plough pan starts at the base of the tilled layer and the depth at which this lies has a greater or lesser effect on crop development. This has a negative effect on crop yields, more pronounced when the compacted layer is only 10 cm deep than when it lies at a depth between 20 to 30 cm (Lowry *et al.*, 1970).

As a consequence of the increased density, there is a corresponding but much more significant increase in the resistance of the soil to root penetration. It was observed by Voorhees *et al.* (1978), when working in a clay-silt loam under the same weight of vehicles, that the soil density increased 20 percent whilst the resistance to root penetration increased more than 400 percent. The values of the resistance to plant root penetration that limit crop development vary from one crop to another.

Cintra (1980) studied the importance for crop systems of changes in soil density, porosity and resistance to root penetration. He observed that soil under natural vegetation, when compared to the same soil under various conventional tillage systems, had greater porosity and less density and less resistance to root penetration. França da Silva (1980) observed a reduction in porosity and an increase in soil density (and in the resistance to root penetration) according to which tillage system was used. The sequential order for these effects for the systems studied was the following: soil under forestry, areas cultivated with animal traction, areas under direct drilling, areas cleared with a tractor mounted rotary slasher and the areas under conventional tillage (Table 11). Cannell and Finney (1973) stated that in general, soil density is higher under direct drilling than under conventional tillage, although this might not always be the case due to the heavier texture or the high organic matter content of the soils.

TABLE 11
Effect of tillage systems on the soil density and porosity
(França da Silva, 1980)

Tillage	Depth (cm)	Soil density (g/cm^3)	Porosity (percent)
Natural vegetation	2 - 15/20	1.01	64.7
	15/20 - 35	1.28	56.6
	35 - 60	1.22	57.9
Tillage with Animal traction	0 - 10	1.04	63.1
	10 - 30	1.18	59.6
	30 - 56	1.21	58.9
Direct seeding	0 - 15	1.40	51.6
	15 - 30	1.41	51.7
	30 - 60	1.22	57.9
Cleared with a tractor-mounted rotary slasher	0 - 15/30	1.33	52.7
	15/30 - 55	1.53	46.1
	55 - 110	1.44	49.6
Conventional tillage	0 - 10	1.17	60.3
	10 - 18	1.44	51.4
	18 - 45	1.22	59.1

The foregoing shows that these indices are useful for the evaluation of the effect of different tillage systems and for identifying the actual physical conditions of the soil.

Soil structure

Soil structure is determined by the arrangement of the elementary particles (sand, silt and clay) into aggregates formed into certain structural models that necessarily include the pore space. Although this might not be considered as a factor that affects plant growth, soil structure exercises an influence on the supply of moisture and air to the roots, on the availability of nutrients, on the penetration and development of roots and on the development of the soil micro-fauna.

From the point of view of soil management, a good structural quality of the soil means a good quality of the pore space or more specifically, good porosity and a good distribution of the sizes of the pore spaces. It follows that the infiltration rate of water, together with the root distribution through the profile are the best indicators of structural quality of the soil (Cabedá, 1984).

The size and the stability of the aggregates can be indicative of the effects of the tillage system and of the crop on the soil structure. Well aggregated soils provide better moisture retention, adequate aeration, easy penetration for the roots and good permeability.

The distribution of the aggregate sizes is one of the important factors in crop development. According to Larson (1964), the aggregates should be reduced to a size which approaches that of the seeds and of the roots of the new plants with the objective of providing sufficient moisture and perfect contact between the soil, the seed and the roots. However, the aggregates should not be so small to the point of favouring the formation of compacted surface crusts and caps. For Kohnke (1968), the ideal size for the aggregates is between 0.50 and 2.00 mm in diameter. Larger aggregates restrict the soil volume explored by the roots and smaller aggregates give rise to very small pores, which do not drain through the action of the forces of gravity. Break-up of the soil aggregates is caused through the intense disturbance of the soil during tillage operations, through a reduction in organic matter content and through the impact of rain droplets on an unprotected surface.

The soil moisture content at the time of tillage is a factor that determines the intensity of the break-up of the aggregates that will occur. The prejudicial effects of the weight of agricultural machinery and of excessive soil tillage under soil moisture conditions that are unfavourable, tend to be cumulative and to intensify with the sequence of annual tillage operations.

Soil disintegration can be reduced by doing less tillage, through the use of crop rotations and through the protection of the soil surface with crop residues. In this way, pastures better assist soil aggregation, followed by direct drilling and by conventional tillage.

Infiltration rate of water in the soil

The infiltration rate of water into the soil determines the rapidity of this process and thus, the volume of water that will remain to run off over the soil surface. When the infiltration rate is low, moisture availability in the root zone can be limiting. The infiltration rate is conditioned by the state of the soil surface, the rate of transmission of the water through the soil profile, the soil moisture storage capacity and flow characteristics of the water. Water infiltration in the soil reflects the soil physical properties. Cropping and tillage systems influence the eventual infiltration rate, both through modification to the roughness and the cover of the soil surface and through changes to the structure, density and porosity of the soil.

Soil tillage can initially improve infiltration and also, sometimes, benefit drainage. But as time passes, tillage favours degradation of the structure and a reduction of the infiltration rate.

Chapter 8

Principal tillage methods

TERMINOLOGY, DEFINITIONS AND CLASSIFICATION OF TILLAGE SYSTEMS

There is confusion in the literature concerning the terminology of tillage because many of the terms used are very general and because there are a very large number of different systems that vary in terms of the implements, the combinations of implements and the intensity of the tillage. Furthermore, different authors often use the same terms for different systems.

The greatest confusion exists between the terms conservation tillage, reduced tillage and minimum tillage.

Conservation tillage is a general term which has been defined as "whatever sequence of tillage operations that reduces the losses of soil and water, when compared to conventional tillage" (Lal, 1995). Normally this refers to a tillage system which does not invert the soil and which retains crop residues on the surface. The percentages of residues that remain after various systems of tillage have been practised are presented in Table 12. Another definition of conservation tillage used is "whichever tillage or sowing system which maintains at least 30 percent of the soil surface covered with residues after sowing so as to reduce erosion by water" (Unger *et al.*, 1995).

However, in some situations and particularly in semi-arid zones, there are insufficient residues or other materials to provide a protective cover to the soil. This may be due to the low straw production owing to the climate, or because they are used for other purposes such as fodder, or are consumed by termites. In this situation the losses of soil moisture can be reduced in comparison to those incurred under conventional tillage systems, by shaping the surface into structures such as ridges and furrows. The ridged tillage system may be considered as a conservation tillage system (Lal, 1995), even though it would be more consistent if there were to be a cover of residues.

Using the first definition, conservation tillage includes the following systems:

- Zero tillage (synonymous with direct drilling and No Till). This refers to planting the seed into the stubble of the previous crop without any previous tillage or soil disturbance, except that which is necessary to place the seed at the desired depth. Weed control relies heavily on the use of herbicides.

R. Barber, Consultant
Food and Agriculture Organization of the United Nations (FAO)
Rome, Italy

TABLE 12
Quantity of residues remaining on the soil after different tillage treatments (Steiner *et al.*, 1994)

Implement	Residue coverage (%)	
	Non-fragile crops[1]	Fragile crops[2]
Mouldboard plough	0-10	0-5
Disc plough	10-20	5-15
Subsoiler	70-90	60-80
Chisel plough with points	60-80	40-60
Stubble mulch chisel plough with points	50-70	30-40
Tandem disc harrow, heavy	25-50	10-25
One-way disc plough, discs of 30-40 cm	40-50	20-40
Field cultivator with duck's foot points	60-70	35-50
Disc harrow followed by land leveller	50-70	30-50
Rotary plough, secondary tillage to 8 cm depth	40-60	20-40

[1] Non-fragile crops include: barley, wheat, maize, cotton, oats, pasture, rice, and sorghum
[2] Fragile crops include: beans, cover crops, groundnut, potato, cardamom, soybean, sunflower, vegetables.

- Strip tillage or zonal tillage refers to a system where strips 5 to 20 cm in width are prepared to receive the seed whilst the soil along the intervening bands is not disturbed and remains covered with residues. The system causes more soil disturbance and provides less cover along the rows than zero tillage.

- Tined tillage or vertical tillage refers to a system where the land is prepared with implements which do not invert the soil and which cause little compaction. For this reason, the surface normally remains with a good cover of residues on the surface in excess of 30 percent. The most commonly used implements are the stubble mulch chisel plough, the stubble mulch cultivator and the vibro-cultivator.

- Ridge tillage is the system of ridges and furrows. The ridges may be narrow or wide and the furrows can be parallel to the contour lines or constructed with a slight slope, depending on whether the objective is to conserve moisture or to drain excess moisture. The ridges can be semi-permanent or be constructed each year, which will govern the amount of residue material that remains on the surface. In the semi-permanent systems which have a good residue cover between the ridges, there will still be more soil disturbance and less overall cover than for the zero tillage system. Generally speaking, this system is less conservationist than strip tillage.

- Reduced tillage refers to tilling the whole soil surface but eliminating one or more of the operations that would otherwise be done with a conventional tillage system. It refers to a wide range of systems such as, for example:
  - Disc harrow followed by sowing;
  - Chisel plough or cultivator, followed by sowing;
  - Rotary cultivator followed by sowing.

Depending on the implements which are used and on the number of passes, reduced tillage can be classified as either a conservation or non-conservation tillage system, depending upon the amount of crop residue which remains after the seed has been placed. It follows that not all reduced tillage systems are conservation systems. Of the three examples cited above, it is probable that only land preparation with the chisel plough or tined cultivator followed by sowing could be classified as a conservation tillage system.

Minimum tillage is the concept that has caused most confusion. It has been defined as "the minimum soil disturbance needed for crop production..."; but the minimum tillage to produce crops varies from zero up to a range of primary and secondary tillage operations which depend on the crop and the type of soil. Sometimes, this term signifies strip tillage or ploughing at the end of the rainy season (the "stale-bed" system). For some authors the term is synonymous with conservation tillage, for others it means zero tillage, whilst for yet others it means reduced tillage. So as to avoid confusion, it is suggested that the term minimum tillage should not be used.

Conventional tillage involves inversion of the soil, normally with a mouldboard or a disc plough as the primary tillage operation, followed by secondary tillage with a disc harrow. The main objective of the primary tillage is weed control through burying, and the main objective of the secondary tillage is to break down the aggregates and to prepare a seedbed. Subsequent weed control may be carried out either mechanically with a cultivator, or with herbicides. The negative aspect of this system is that the soil lacks a protective residue cover and is left practically bare, meaning that it is susceptible to soil and water losses through erosive processes.

One way to visualize the tillage terminology is to imagine a triangle (see Figure 12). Along the left side is found conventional tillage, which includes the complete set of operations for land preparation. As the triangle becomes narrower towards the right, the number of tillage operations reduces, corresponding to reduced tillage. At the right of the triangle, soil tillage is eliminated completely representing zero tillage. One may also classify tillage on the basis of the degree of soil disturbance and on the residue cover which remains after sowing (Table 13).

TABLE 13

Tillage systems classified according to the degree of disturbance to the soil and the surface cover of residues

Conventional tillage		Non-Conservation tillage	Conservation tillage				
Mouldboard plough	Disc plough	Reduced tillage	Reduced tillage	Ridge tillage	Tined tillage	Strip tillage	Zero tillage
---→ Less soil disturbance both in intensity and frequency							
---→ Increased cover with residues							

PRINCIPAL TYPES OF TILLAGE SYSTEM

The most important tillage systems will be dealt with in more detail below. These comprise conventional tillage, reduced tillage, ridge tillage, tined or vertical tillage, strip tillage, zero tillage and combined systems of tilling and seeding and deep tillage or subsoiling.

Conventional tillage

The principle of conventional tillage is based on soil inversion with the objective to control the weeds, followed by various operations for preparation of the seedbed.

Advantages

- Provides good weed control with low herbicide costs;

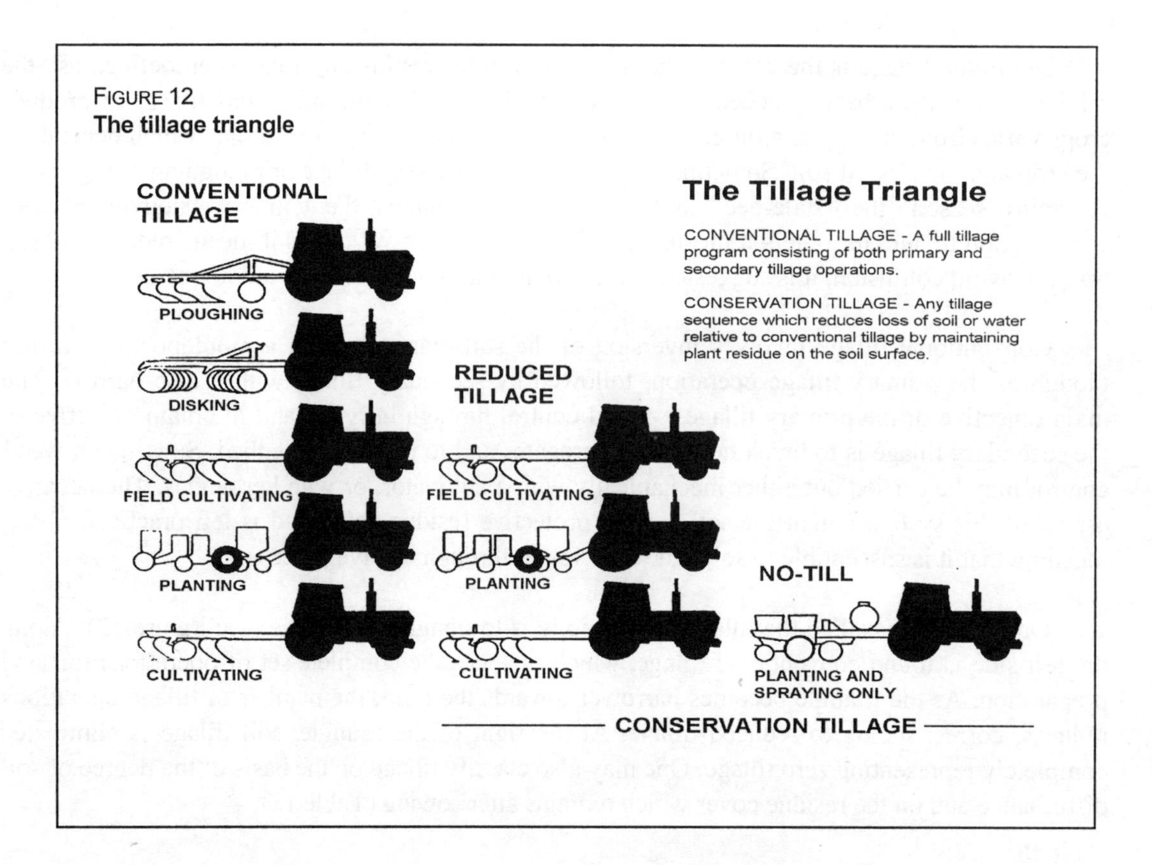

FIGURE 12
The tillage triangle

- Allows the control of diseases and insects by burying the crop residues;
- Eases the operations of incorporating fertilizer, lime, pesticides and pre-sowing herbicides;
- Eases loosening of the soil surface, compacted layers and crusts;
- Is adapted to the incorporation of pastures in crop rotation systems;
- After a single ploughing operation, it creates a rough surface that improves the infiltration of rainwater.

Limitations

- The soil is left bare and thus, susceptible to crust formation and erosion due to water or wind;
- Requires much equipment for the various different operations;
- Sometimes, in order to save time, large and heavy tractors are used which increase soil compaction;
- Increased fuel consumption, takes longer to prepare the seedbed and is less flexible when timeliness of sowing is important due to the climate;
- The subsurface soil can eventually come up to the surface which, if the physical and chemical characteristics of the subsoil are unfavourable, could cause problems of seed germination and during the initial growth of the plant;

- The inversion and the various tillage operations produce a loose soil which is susceptible to compaction;
- The base of the plough smoothes the soil causing blockage of the pores which hinders the permeability of the surface soil;
- Ploughing every year at the same depth forms a compacted layer at the foot of the plough. This is common when the soil surface is dry but the moisture content at 20 cm depth is high;
- The elevated number of tillage passes for seedbed preparation results in a loss of soil moisture. Although when commencing tillage the soil might have an appropriate moisture content for seed germination, once the seedbed is completed the soil could be too dry for sowing. In this case it would be necessary to await the next rainfall before being able to sow.

Machinery

Primary tillage requires the use of either a mouldboard or a disc plough. A reversible mouldboard plough increases working efficiency. After primary tillage, a disc harrow and occasionally a spike-toothed harrow is needed. A conventional seed drill is then used and weed control requires a sprayer or a row-crop cultivator, or both.

Operations

The mouldboard or disc plough inverts the top layer of soil, normally down to a depth of some 30 cm. Afterwards, various passes are made with a disc harrow until the aggregates are broken down to a size appropriate for the seedbed, the number of passes depending on the soil texture and moisture content. As a general guide for avoiding the formation of surface crusts, the soil preparation should leave aggregates the approximate size of an orange (6 to 8 cm in diameter) for light and medium soils, and aggregates the size of eggs (4 to 5 cm in diameter) for heavy soils.

If it is necessary to level the land, a spike-toothed harrow is used. It is not advisable to use a wooden plank in light to medium soils, as this will tend to pulverize the soil; such a tool however, can be useful in heavier soils. It would be better to use a levelling blade mounted on the disc harrow to achieve good levelling.

In many crops a pre-sowing herbicide is applied and incorporated during the last pass with the disc harrow or the spike-toothed harrow. A conventional seed drill is used to place the seed and this should be equipped with press-wheels rather than chains to cover the seed. If the seed drill has seed coverage chains it is necessary to prepare a finer seedbed in order to achieve good germination. However this finely structured soil is more susceptible to crust formation and to erosion by water.

Reduced tillage

The term “reduced tillage” refers to tillage systems where there is less frequency and intensity of cultivation as compared to conventional systems. This definition is extremely broad and it follows that tillage systems which vary as regards the implements used, the frequency and the intensity of operations, can all be considered as reduced tillage systems. The type of implement and the number of passes also vary; the result is that some systems leave very little surface residue and in others, this may be in excess of 30 percent. For this reason, some reduced tillage systems are classed as a type of conservation tillage whilst others are not. Generally, reduced tillage systems do not use either mouldboard or disc ploughs.

Owing to the great variation in reduced tillage systems, it is difficult to generalize over the advantages and limitations. However, all the systems have the advantage of reducing fuel consumption, work time and the equipment required as compared to conventional tillage. Reduced tillage systems are thus more flexible than conventional systems. Germination conditions tend to be generally better than for zero tillage systems due to the break-up of the soil; there is also comparatively more flexibility for weed control using cultivators or herbicides.

Three systems of reduced tillage will be briefly discussed below:

With the disc harrow

One or two passes with a disc harrow are made for this system. Sowing is then done normally with a conventional seed drill. The advantages are the savings in fuel and time and the creation of favourable seed germination conditions.

The limitations are that often, only a very scant cover of residues remains on the surface, although this depends on the angle set for the discs on the harrow and the number of passes. The greater the disc angle, the greater is the soil disturbance and the less the surface cover which remains. In this case, the soils are left susceptible to crust formation. In soils susceptible to compaction, the various passes made each year with a disc harrow working to the same depth (typically from 10 to 12 cm), result in the formation of a hard pan at this depth. Shallow tillage makes weed control more difficult and hence it is normally necessary to resort to the use of herbicides.

With chisel plough or field cultivator

This system comprises two passes with the chisel plough or the field cultivator, followed by seeding. A single pass with the chisel plough is normally insufficient to loosen the soil. The advantages are as previously mentioned but in addition, the infiltration rate of rainwater is improved, particularly in those soils that are susceptible to compaction and hardsetting. Normally with this system, a residue cover remains over more than 30 percent of the surface area for which reason the system is considered conservationist and affords protection of the soil against erosion.

The limitations are that the physical conditions of the soil and the surface undulations make sowing difficult and thus also, the seed germination. It is also difficult to uniformly incorporate the pre-seeding herbicides.

With rotary cultivator (rotavator)

This system has the same advantages of those mentioned above. The main limitation is that the rotary cultivator tends to pulverize the soil and leaves very little residue cover on the surface, which thus is left in a condition susceptible to crust formation. In addition, the system can cause the formation of a plough pan over time.

Tined cultivation or "vertical tillage"

Introduction

The following section has been taken directly from the article titled "Labranza Vertical" by Barber *et al* (1993), a term that can be directly translated from the Spanish as "Vertical tillage" but which is more commonly referred to as "Tined tillage". The publication was written for the farmers, soils and climate of the area of Santa Cruz, Bolivia. This area is situated in the subhumid tropics and is characterized by two rainy seasons and by two annual crops. During the second, winter season, there is less rainfall and temperatures are lower than during the first, summer season.

Tined cultivation is characterized by using rigid or flexible tines equipped with points in the place of discs, so as to loosen the soil without causing soil inversion. This leaves a protective cover of residues of the previous crop and uprooted weeds on the soil surface.

The principal implements used in tined tillage consist of the stubble mulch chisel plough (Figure 13), the spring-tined cultivator (Figure 14) and the stubble mulch cultivator (Figure 15). It should be noted that the terminology employed for the implements illustrated follows manufacturers trade descriptions or generally recognized terminology. The illustrations serve to clarify the definitions employed.

FIGURE 13
Stubble mulch chisel plough

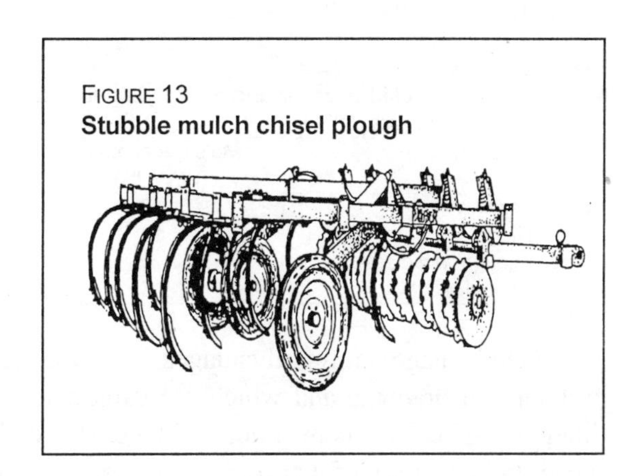

FIGURE 14
Spring-tined cultivator (Vibro-Cultivator) equipped with a levelling blade and a long-fingered rake

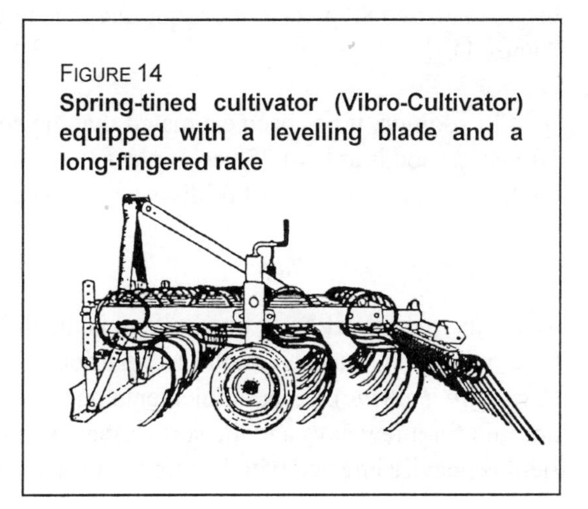

FIGURE 15
Stubble mulch cultivator

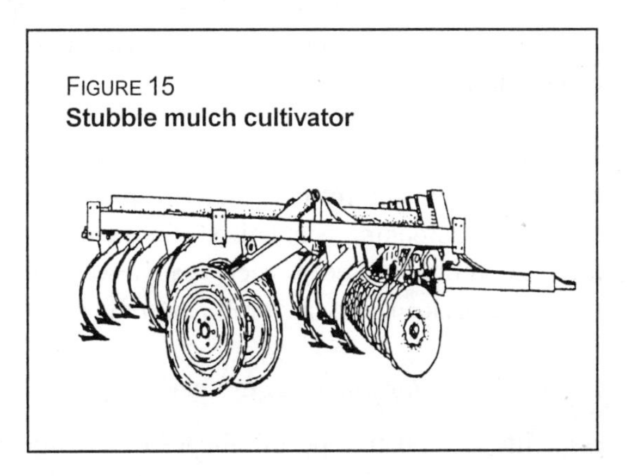

Advantages of tined tillage

Tined tillage better sustains the soil productivity because it leaves residues on the surface that protect the soil against the processes of erosion. This residue cover also impedes the formation of surface crusts that could cause reduced crop emergence.

Tined implements cause little soil compaction at depth, which would otherwise limit root penetration. In contrast, the discs of conventional tillage equipment cause the formation of a plough pan.

Because tined tillage does not invert the soil, there is less decomposition ofthe organic matter and less loss of moisture, which is very important before sowing. Table 14 shows an example of the effects of tined tillage on soil moisture content, residue cover and the yield of maize, as compared with other tillage systems.

TABLE 14
Moisture content, residue cover and maize yield for four tillage systems in Oxford, North Carolina, USA in 1985.
(Source: Cook and Lewis, 1989)

Tillage system	Moisture content (W/W, %)	Residue cover (%)	Maize yield (t/ha)
Zero tillage	13	90	5.77
Tined tillage	12	33	5.58
Chisel plough with discs	9	14	4.70
Conventional tillage	6	3	3.57

Tined tillage shows advantages in a wide range of soil types, including those which have problems of drainage and which are subject to compaction. The operational efficiency of tined tillage is higher than conventional tillage, above all because the vibro-cultivator works at a higher speed and has a wider working width than the disc harrow. Because of this, tined tillage allows between 50 and 80 percent more surface area to be prepared per day than conventional tillage (see Annex 1).

In addition, it has been estimated that the cost to acquire and maintain tined equipment over a 10-year period is at least 25 percent less than conventional tillage equipment. This is mainly due to the higher maintenance cost of discs as compared to the tine points (see Annex 2).

Limitations of tined tillage

The principal limitation of tined tillage is the difficulty to mechanically control stolon and rhizome type weeds in humid conditions, particularly graminaceous weeds such as Bermuda grass (*Cynodon dactylon*). Tined implements pull out the weeds and leave them on the surface. If it does not rain for a few days and the soil surface is moist, they sprout again easily. For this reason, if the field is heavily infested with Bermuda grass it is better not to use tined implements.

This problem is much more serious for crops such as maize and sorghum where there are no herbicides available for the post-emergent control of graminaceous weeds (or they are uneconomic).

Another limitation of tined tillage is that it could result in increases in pests and diseases associated with the residues that are not completely buried. This would happen more probably where monocropping is practised (for example soybean followed by soybean each year), or under the same rotation of crops each year (for example a soybean-wheat rotation). However until now, there is no evidence of this problem in the area of Santa Cruz.

Pre-requisites for introducing tined tillage

As with any other type of tillage, tined tillage gives best results in fertile soils that are not compacted, are well drained and level, and do not have weed problems. If the soil is compacted, these layers must be broken up, the field levelled if necessary and any nutritional deficits rectified before introducing the practice of tined tillage.

It is also advisable to select fields that are not infested with graminaceous weeds and to start with summer crops of soybean and winter crops of wheat, soybean or sunflower. In the case of wheat, generally there are no undue problems with graminaceous weeds, and with soybean or sunflower it is easy to control graminaceous weeds with post-emergent herbicides.

For the successful introduction of tined tillage it is important that the residues and the weeds be well chopped and uniformly spread over the plot. This will avoid blockages of the implements and so combine harvesters should be equipped with a straw chopper and spreader.

Furthermore, a rotary mower is necessary to reduce the length of the stalks of maize, sorghum, sunflower and cotton after these crops have been harvested. Neither should weeds be allowed to develop when the field is left fallow. They should be mown as soon as they reach a height of 15 cm to avoid weed competition problems and to avoid implement blockages.

Primary tillage for summer crops

The use of a stubble mulch chisel plough (Figure 13) is recommended, which consists of a chisel plough equipped with a front-mounted gang of cutting discs and straight points with a width of approximately 4 to 5 cm. The tines should be fixed across four crossbeams to reduce possibilities of blockages and the tractor should be driven at a speed of between 6 and 9 km/h.

The tine spacing will influence the working depth, the number of passes needed and the power requirement of the tractor. As a general rule, the depth of work should be equal to the tine spacing divided by 1.1 in order to obtain good soil loosening across the entire working width of the implement. If the tines are more widely spaced than about 30 cm, it will be necessary to make two overlapping passes. In addition, once the working depth exceeds 18 cm, normally two passes will be necessary, depending on the soil texture and moisture content. Tractor power requirements are generally in the range of from 9 to 12 hp per tine. Recommendations concerning these specifications are summarized in Table 15.

TABLE 15
Working characteristics of the chisel plough for stubble. (Source: Barber *et al.*, 1993)

Tine spacing (cm)	Number of tines	Working depth (cm)	Number of passes	Working width (m)	Minimum tractor power (HP)
21	11	18-20	1	2.31	110
21	13	18-20	1	2.73	130
26-28	9	17-18	2*	2.43	90
26-28	9	24-26	1-2	2.43	90
26-28	11	17-18	2*	2.97	110
26-28	11	24-26	1-2	2.97	110
26-28	13	17-18	2*	3.51	130
26-28	13	24-26	1-2	3.51	130
35	7	18-20	2*	2.45	70
35	9	18-20	2*	3.15	90
35	11	18-20	2*	3.85	110
35	13	18-20	2*	4.55	130

* Crossed

The stubble mulch chisel plough is recommended when the soil is friable, which corresponds to a moisture content between dry and slightly moist. In this condition, the soil disintegrates easily in the hands and the chisels break up the clods through their vibratory action. This will also control the weeds. If the soil moisture content is too high, the soil is more plastic and the chisels will produce fissures without breaking up the clods and without controlling the weeds. In contrast, if the soil is too dry, the clods will be very resistant and again will not be broken up.

The first pass should be made as soon as possible after the harvest and in any case, before the weeds have grown to a height of 15 cm. This reduces the risks of weed infestation and of implement blockages.

In case a second pass is necessary in order to reach the required working depth and to achieve good soil loosening and uprooting of the weeds, it is preferable to make it across the first pass at 90°, and never at an angle less than 30°. However, on many occasions this is not feasible because the plots are too long and narrow.

The second pass may be done the same day if the moisture content is optimum, but if the soil is rather moist it will be necessary to wait a few days for it to dry out.

If the soil is of light or medium texture with symptoms ofcompaction beginning to show, it is advisable that the second pass with the stubble mulch chisel plough be at a depth of at least 25 cm. It is important to note that the stubble mulch chisel plough is not a subsoiler. If the soil is truly compacted, various passes with a subsoiler will be needed to break up the hard pan. In this situation, primary tillage with the stubble mulch chisel plough will not be necessary and one may proceed directly to the secondary tillage operations.

Use of other implements for primary tillage

Instead of using a stubble mulch chisel plough, one may use a flexible tined chisel plough such as the "Vibroflex" (Figure 16), which has vibrating tines and straight points with a width of 6.5 cm. The points are spaced at about 23 cm and distributed across four beams. This implement works like a chisel plough but works more rapidly due to it being lighter and because of the vibration of the tines.

FIGURE 16
The flexible tined chisel plough "Vibroflex"

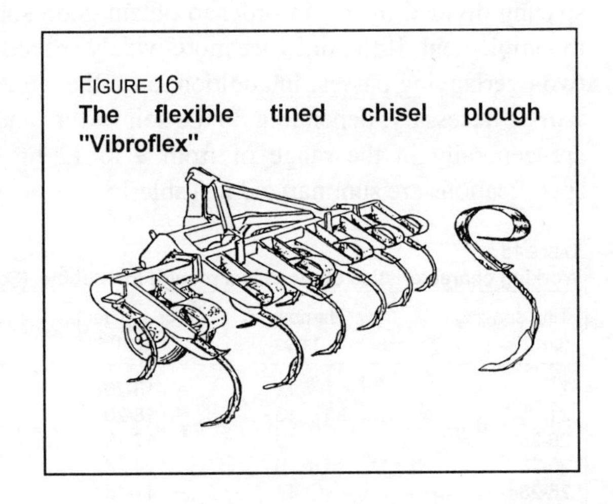

The vibrations also favour breaking up the clods and shake the soil off the roots of the weeds. In light and medium soils, the power requirement is about 6 hp per tine. Optimum work quality is achieved by operating at a speed of from 8 to 12 km/h which means that the overall operational efficiency is superior to that of the stubble mulch chisel plough. However, due to the absence of the front gang of cutting discs, one may encounter problems of blockages when there are heavy weed infestations or large quantities of residues.

If a stubble mulch chisel plough is not available but there is a normal chisel plough that is not equipped with a front-mounted gang of cutting discs, it is probable that blockages will occur, particularly when there are abundant weeds and residues. It will only be advisable to use this common chisel plough when there are few residues and weeds (particularly creepers) and when the soil is friable. In addition, the tractor should work at a speed of between 7 and 9 km/h. A good option could be to modify a chisel plough, adding the cutting discs and pressure springs.

Secondary tillage for summer crops

The objective of secondary tillage is to prepare the seedbed, breaking up the biggest clods, pulling out the weeds, levelling the surface and leaving most of the residues on the surface. It also serves to incorporate any pre-seeding herbicides.

A spring-tined cultivator such as the Vibro-cultivator (Figure 14) is recommended for secondary tillage. This replaces the lightweight disc harrow and has several other advantages for seedbed preparation. The Vibro-cultivator has flexible and vibrating spring tines, spaced at about 10 cm and mounted on four crossbeams. In case there are heavy residues or weeds, it is recommended to increase the tine spacing to 15 cm to reduce the risk of blockages. Straight points about 3.5 cm wide should be fitted or, if there is not much residue cover, double points (Figure 17).

FIGURE 17
Types of cultivator point

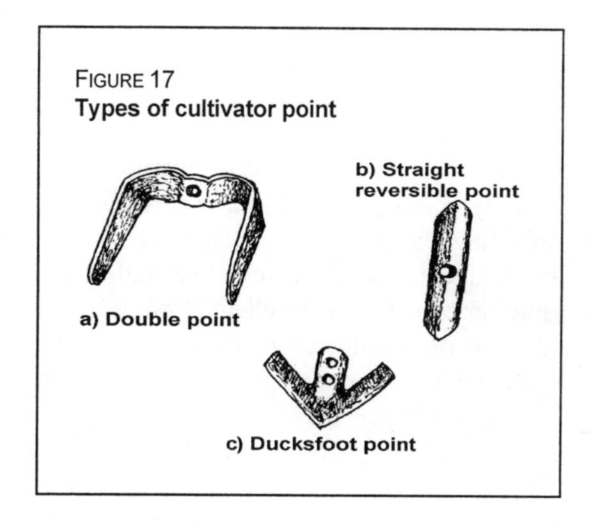

Working depth should be between 8 and 10 cm. It is very important to maintain a working speed of 8-12 km/h to optimize the vibrations, which shake the weeds and break up the clods. The vibratory action leaves the larger clods on the surface, which better resist the formation of crusts and leave the smaller aggregates in the lower levels, so assisting seed germination.

The Vibro-cultivator works well in friable soils, causing good clod disintegration but for larger clods and in dry conditions, little clod break-up is achieved. Under these conditions it could be necessary to make one pass with a lightweight disc harrow with a disc diameter of less than 55 cm (22 inches) to break down the clods.

One or two crosses with the Vibro-cultivator are recommended, the number depending on the weed control required and the size of the clods. To prepare a good seedbed in light or medium soils, the aggregates should have a size of between 6 and 8 cm in diameter. In contrast for heavy soils, they should be broken down to a diameter of only 4 or 5 cm. In this way, a seedbed is achieved that will reduce the risks of crust formation and will assist seed germination. To reduce the clod size in light and medium soils, it is recommended that the secondary tillage be undertaken when the soil is friable. It is not recommended to use either cage-type rolls or helicoidal type rolls as they cause pulverization of these soils.

In order to level the soil, it is recommended that a long-fingered rake (Figure 18) or a spike-toothed harrow (Figure 19) be coupled behind the Vibro-cultivator. But if there is an excessive residue cover, coupling the rake will cause the residues to be dragged across the surface. If the surface irregularities are not too pronounced and do not affect the seeding operation, it would be better not to undertake levelling. In contrast, if the surface is very uneven or ridged after primary tillage and the Vibro-cultivator cannot level it properly, it would be advisable to level it by making a pass with the lightweight disc harrow with discs of less than 55 cm in diameter. This should be done when the soil is well dried.

Coupling a levelling blade to the Vibro-cultivator (Figure 14) is not recommended for light or medium soils as it tends to pulverize the soil. It is however, recommended for moderately heavy soils.

For reduction of the clod size in moderately heavy and heavy soils, it is recommended that one or two lightweight rolls be coupled to the Vibro-cultivator of the cage type with angled bars (Figure 20). Adjusting the pressure on the rolls controls the degree of clod disintegration and at the same time the land is levelled. However, with this type of roll, due to the angled bars, there is partial incorporation of the residues. For this reason, one may use a heavy clod-breaking type of roll with helicoidal bars (Figure 21), coupled behind the Vibro-cultivator. Alternatively, the roll can be pulled directly behind the tractor. If soil moisture conditions are optimum, this roll breaks up the soil aggregates, levels the surface and does not incorporate the surface residues.

FIGURE 18
Long-fingered rake

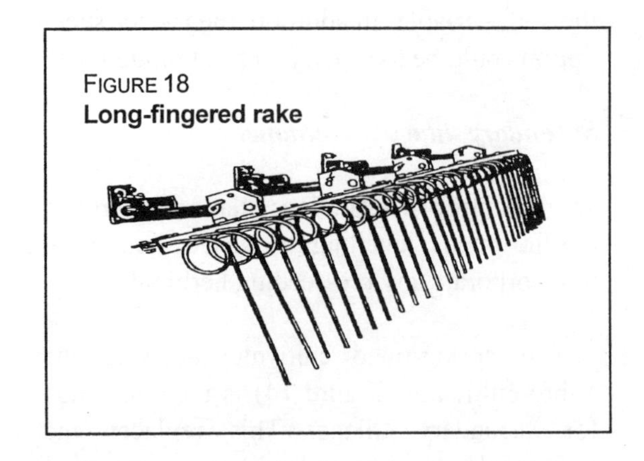

FIGURE 19
Spike-toothed harrow

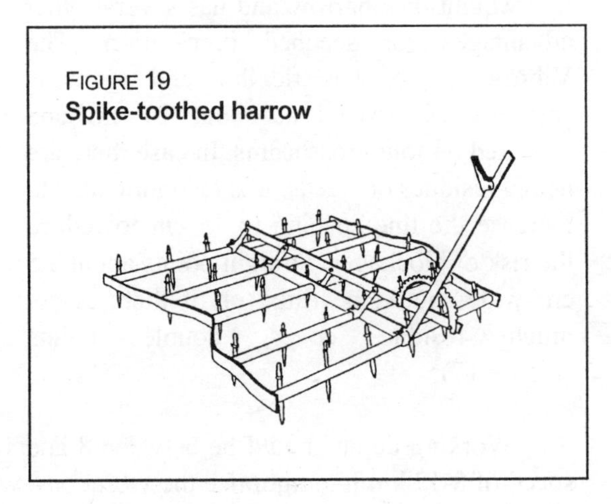

FIGURE 20
Lightweight cage rolls

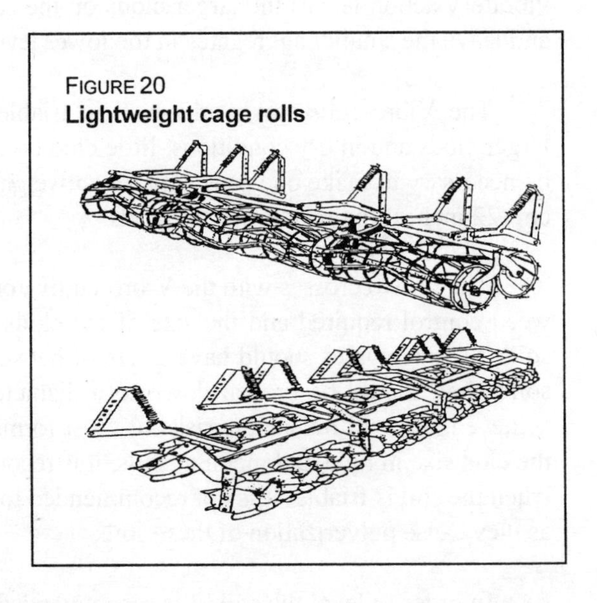

Use of other secondary tillage implements

If a Vibro-cultivator is not available, a field cultivator may be used, preferably a stubble mulch cultivator which has a front-mounted gang of cutting discs (Figure 15), or a flexible tined chisel plough (Figure 16). These implements should be fitted with duck's-foot points, about 25 cm wide (Figure 17). The tines should be spaced at between 17 and 22 cm and mounted alternately along four crossbeams to reduce residue blockages. The speed of work should be maintained at between 8 and 10 km/h and the power requirement will be about 5 hp per tine. It is only recommended to use these implements when the soil is dry and to work down to a depth of 7 to 9 cm. If the soil is slightly moist, the duck's-foot points can cause soil compaction.

Alternatively, a lightweight disc harrow can be used for secondary tillage, although this is not the most recommended method as it incorporates most of the surface residue, leaving the surface bare. It also causes a loss of soil moisture and compaction.

Neither is it recommended to attach a wooden plank behind the harrow to level light or medium soils as this causes strong pulverization of the superficial soil, favouring crust formation and erosion.

FIGURE 21
Heavy clod-breaking roll

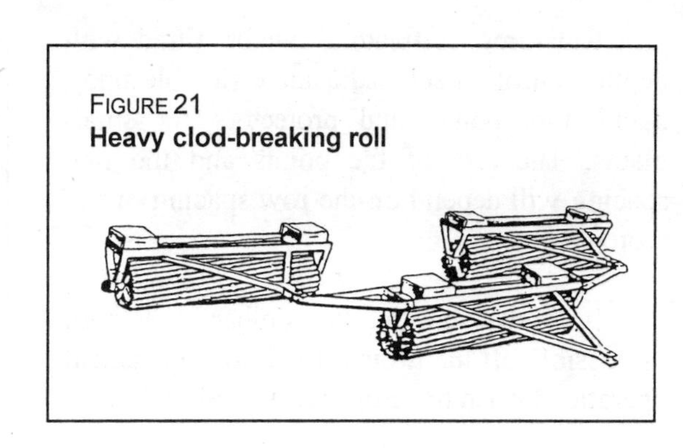

Tined tillage for winter crops

Seedbed preparation using tined tillage for winter crops should be similar to that for summer crops but with less passes and at a shallower working depth, except in the case that the soil has become excessively compacted over the summer.

In light to medium soils when the first 10 cm of the soil are dry, two passes are recommended with a stubble mulch cultivator equipped with duck's-foot points about 25 cm wide and working down to a depth of from 7 to 9 cm. After this, the soil should be ready for sowing. If however the soil is rather moist in the surface layer, one pass with a stubble mulch chisel plough is recommended with straight points and working down to a depth of between 10 and 15 cm.

Afterwards, one or two passes should be made with the Vibro-cultivator with the tines spaced at 15 cm and working down to a depth of 8-10 cm. The number of passes will depend on the size of the clods and the degree of weed control required. Furthermore, in order to level the surface, a long-fingered rake or a spike-toothed harrow should be coupled behind the field cultivator or Vibro-cultivator.

For heavy soils, two passes are recommended with the stubble mulch chisel plough equipped with straight points and working down to a depth of between 10 and 15 cm. Following this, one pass should be made with the Vibro-cultivator fitted with straight points and working to a depth of 8 to 10 cm. A double helicoidal type roll or cage roll should be coupled behind the vibro-cultivator. After completing these operations, the land is ready for sowing.

Weed control in tined tillage

It is recommended to apply pre-seeding herbicides, incorporating them with the Vibro-cultivator. For the control of post-emergence weeds, it is preferable to apply herbicides with the objective of reducing the risks of soil compaction that could occur under moist conditions, through the use of row-crop cultivators.

When the soil is dry, row-crop cultivators do not cause compaction and they can be combined with band sprayers (Figure 22). This type of sprayer applies the herbicide in narrow strips, only along the length of the rows of crops and so saves approximately 50 percent of the herbicide.

Row-crop cultivators can be fitted with depth control wheels, adjustable flexible tines, duck's-foot points and protectors for small plants. The size of the points and the tine spacing will depend on the row spacing of the crop.

It is advisable that the toolbar be as high as possible off the ground to allow mechanical operations when the crops are already tall.

The row-crop cultivators should work at speeds of between 8 and 12 km/h and to a depth of from 5 to 8 cm. If a row-crop cultivator is not available, a chisel plough or a Vibro-cultivator can be adapted. In the latter case, longer tines will need to be fitted.

FIGURE 22
Row-crop cultivator equipped with a band sprayer

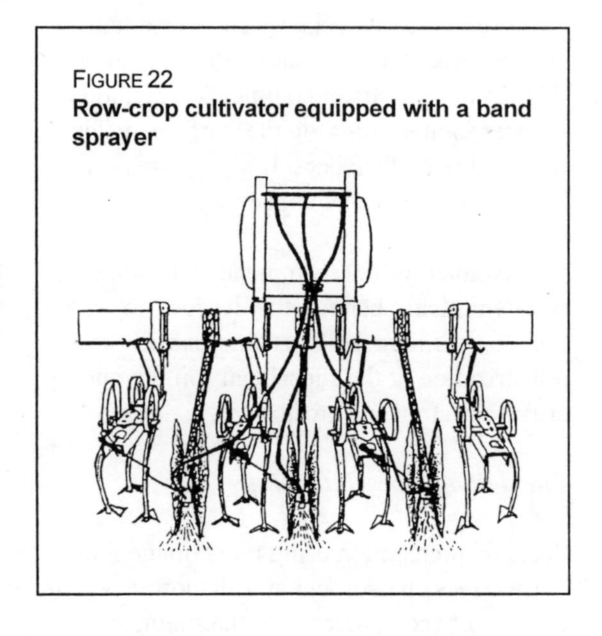

Use of tined equipment after land clearance

It is advisable to use a system of conventional tillage with the "Rome Plow" (a fairly heavy disc harrow), working down to a maximum depth of 15 to 18 cm to level the land. This should only be used during the first or during a maximum of the first two seasons immediately after the land clearance operation.

During this period it is important to remove all roots from within the topsoil. A few passes down to 15 or 20 cm depth with a ripper, which is an implement similar to a subsoiler but which does a more superficial job (Figure 23) will help remove the superficial roots.

FIGURE 23
Ripper

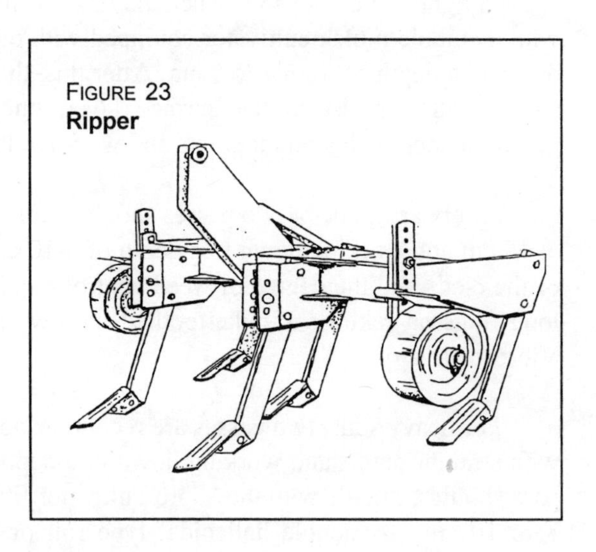

Afterwards one can start with the tined tillage system, working to a depth no greater than 12 to 15 cm over the first few years until the deeper roots have decomposed. In this way, the risks of breaking the tines and the cutting discs of the tined equipment are reduced. Also, the roots which are present deeper than 15 cm serve as organic manure and for this reason it is not advisable to remove them.

Sowing after tined tillage

Conventional seed drills can normally be used after land preparation with tined equipment. Because the amount of surface residue remaining, when sowing takes place, is not generally large due to the rapid rate of decomposition, it is unlikely that this will adversely affect the sowing operation.

In order to ensure good contact between the seed and the soil, it is advisable to use pressure wheels at the back of the seed drill rather than covering chains.

Conclusions

a. Tined tillage is more conservationist, efficient and economic than conventional tillage:

b. Tined tillage is appropriate for a wide range of soil types, but it is not advisable to use the system in heavily weed-infested areas, and those which have a lot of Bermuda grass (*Cynodon dactylon).*

c. It is important that the combine harvester be equipped with a straw chopper and spreader and that the straw and weeds are mown to ensure a uniform distribution of the residues. In this manner, residue blockages in the equipment will be avoided.

d. One or two passes with the stubble mulch chisel plough are recommended, working down to a depth that depends on the tine spacing used during the primary tillage operation. The first pass should be done as soon as possible after the harvest is completed and the tractor should travel at a speed of between 7 and 9 km/h. Alternatively, a spring-tined chisel plough can be used. This travels more quickly than the chisel plough but it does not have a front-mounted gang of cutting discs.

e. It is recommended that secondary tillage consist of one or two passes with the Vibro-cultivator at a depth of 8-10 cm. The tractor should work at a speed of between 8 and 12 km/h.

f. Levelling of light to medium soils should be done with a long-fingered rake or a spike-toothed harrow. The use of cage or helicoidal rolls is not recommended for soils of these textures.

g. For moderately heavy and heavy soils, it is recommended to couple one or two lightweight cage-type rolls or a heavy clod-breaking roll to reduce the clod size and also to level the surface.

h. A field cultivator fitted with duck's-foot points can substitute the Vibro-cultivator. It may also be used for soil preparation of winter crops, but it should not be used when the soil is even slightly moist.

i. After land clearance, it is recommended that a "Rome Plow" be used during the first one or two seasons whilst the roots are removed from the upper 15 cm depth. Afterwards, tined tillage may be introduced, working to a depth of between 12 and 15 cm.

j. After seedbed preparation with tined implements, a conventional seed drill fitted with pressure wheels can be used to sow the crop.

STRIP TILLAGE OR ZONAL TILLAGE

This tillage system consists of preparing strips of seedbed which are only 5 to 20 cm wide and 5 to 10 cm deep. Soil between the strips is not touched except for weed control and the bands remain with a protective cover of dead weeds and stubble.

Advantages

- Loosening the soil along strips gives good seedbed conditions for the germination and initial growth of the plants;
- A conventional seed drill may be used;
- The presence of a protective soil cover between the strips encourages infiltration of the rainfall;
- There are less problems of erosion and of crust formation in the soil between the sown strips;
- Less fuel is expended, the equipment suffers less wear, and the land preparation requires less time;
- There is no need for high-powered tractors;
- It is easy to apply fertilizer along the strips of disturbed soil;
- The system is suitable for compacted and hardsetting soils.

Limitations

- The soil along the strips can form surface crusts that will hinder crop emergence. It is less suitable for soils that are susceptible to crust formation;
- It is difficult to prepare the strips giving good seedbed conditions whilst still using conventional implements. It is better to use specialist equipment, which often is not available (see the *Strip Tillage System* mentioned in the section dealing with combined systems for soil preparation and seeding).

RIDGE TILLAGE

The ridges may be either wide or narrow in this system and the furrows can work in two ways: trapping and collecting the rainwater in semiarid zones or draining off the excess water in wet zones. It follows that the system must be designed according to the specific needs, be they to conserve moisture, to drain excess water or to receive water for gravity fed irrigation systems. The ridges and furrows can be made by hand, with animal traction or with machinery. In addition, the ridges can be made every year or they may be semi-permanent with only maintenance operations being done each year. In the systems when ridges are made every year, very little residue cover remains on the surface, whereas for the semi-permanent systems the soil cover depends on the method for controlling the weeds and managing the residues. There are also broad cambered beds with a width between seven and ten metres.

Advantages

- When the ridges are constructed parallel to the contour they conserve the moisture in semi-arid and sub-humid zones. The rainfall becomes trapped between the furrows where it infiltrates rather than being lost through runoff. The infiltration may be improved by making ties or barriers in the furrows at distances of between one and three metres (see Table 16 for an example of the effect of tied ridges on the yields of different crops in Tanzania).

TABLE 16
Effect of tied ridges on the yields of different crops in Tanzania (Prentice, 1946)

Year	Rainfall (mm)	Crop	Yield (kg/ha)		Difference (%)
			On the flat	On tied ridges	
1939	610	Cotton	323	542	68
1939	610	Sorghum	202	734	263
1940	787	Sorghum	808	1 122	39
1942	1 245	Cotton	1 049	854	-18
1943	585	Maize	172	825	380
1944	660	Cotton	101	393	290
1944	660	Sorghum	853	869	2
1944	660	Sorghum	343	798	133
1945	787	Cotton	684	1 234	80
1945	787	Sorghum (residue)	1 467	3 747	139
1945	787	Sorghum	976	892	-9

When the ridges and furrows are made with a slight slope, they drain off the excess moisture in soils with drainage problems or in humid or very humid zones. This method drains off the excess moisture by the lateral and superficial movement of the water from the ridges towards the furrows. Sowing on the ridges can also have the effect of raising the crop's rooting zone above the impermeable layer or above the water table. This results in better germination and deeper root growth. The system is very suitable for Vertisols and other clay soils with drainage problems.

- The soil on the ridges does not suffer compaction.
- Loosening the soil on the ridges gives better conditions for germination.
- The system of ridges and furrows makes it easier to combine different crops grown in the furrow and on the ridge at the same time.
- The resistance of the crusts that form on narrow ridges is lower on the crest due to the formation of tension cracks that favour crop emergence.

Limitations

- For the systems with the ridges made every year, little protective residue cover remains on the soil and consequently there are greater risks of crust formation and water erosion.
- The system is not suitable for slopes greater than seven percent due to the risks of accumulating an excessive amount of water in the furrows, which cause the collapse of the ridges or spilling over the top of the ridges.
- It requires a great deal of manual labour to construct ridges by hand, and more time when using animal traction or mechanized systems.
- It needs more time for maintenance of the ridges and furrows.
- In semi-permanent ridges, only crops with the same spacing can be sown in mechanized systems.
- The soils become susceptible to erosion for sometime after maintenance work has been carried out to the ridges or after cultivation for weed control.

Zero tillage

Advantages

- Zero tillage reduces the risks of erosion and so may be practised on slopes that are much steeper than would be possible with conventional tillage systems (see Table 17 for a comparison of the effects of zero tillage and conventional tillage on losses of soil and water in Nigeria).

TABLE 17
Effects of tillage on runoff and soil loss for soils cultivated with maize in Nigeria (Source Rockwood and Lal, 1974)

Slope %	Bare tilled soil		Ploughed		Zero tillage	
	Runoff %	Erosion (Mg/ha)	Runoff (%)	Erosion (Mg/ha)	Runoff (%)	Erosion (Mg/ha)
1	18.8	0.2	8.3	0.04	1.2	0.001
5	20.2	3.6	8.8	2.16	1.8	0.001
10	17.5	12.5	9.2	0.39	2.1	0.005
15	21.5	16.0	13.3	3.92	2.2	0.002

- It increases the infiltration rate for the rainfall, reduces evaporation and through this, increases moisture retention in the soil.
- It increases the organic matter content in the topsoil, improving the soil structure.
- It stimulates biological activity, the increased activity of the macro-fauna resulting in greater macro-porosity.
- It reduces the high temperatures and temperature variations in the seed zone.
- It reduces fuel consumption by about 40 to 50 percent owing to the limited number of operations: only one pass is needed for preparation and seeding.
- It reduces the time and labour requirements up to 50 or 60 percent. This is very advantageous during critical periods, particularly when few days are available for sowing the crop. It follows that the system is more flexible than other conventional systems. Sometimes, owing to the short time needed for sowing, two crops a year may be sown instead of only one.
- It reduces the number of machines required, the size of the tractors and the costs for repairs and maintenance of the machinery.
- Frequently, the yields are higher under zero tillage, particularly in zones suffering from a moisture deficit.
- It is suitable for light and medium soils, well-drained soils, volcanic soils and for sub-humid and humid areas.

Limitations

- Zero tillage is not suitable for degraded or badly eroded soils.
- It is not suitable for soils that are very susceptible to compaction or for hardsetting soils because the compacted layers cannot be loosened, which hinders emergence, initial crop development and root growth.

- It is not suitable for poorly drained soils or massive clayey soils unless they are naturally self-mulching due to the difficulties in creating good conditions for germination.
- It is not suitable for land that has recently been cleared and which still has roots in the surface layer, because of the risk of damaging the seed drill.
- It requires good knowledge of weed control methods because it is not possible to correct errors by mechanical control methods.
- There may be an increase in the population of weeds that are more difficult to control.
- It is not suitable for soils that are infested with weeds due to problems in their control.
- It requires special and costly machinery.
- It is difficult to incorporate pesticides against soil-borne insects and phosphate fertilizers, which have to be placed into the soil.
- In order to modify a direct seed drill so that it can place fertilizers into the soil, it is necessary to add additional units of cutting discs and disc openers to avoid clogging from residues.
- There may be additional problems with diseases and pests due to the persistence of the residues on the soil surface, which creates a better environment for their development. However, the presence of residues can also stimulate the proliferation of the natural predators of the pests. It is very important to regularly check so as to control the incidence of pests. In the case of cotton, there may be additional pest problems because it is not possible to bury the crop residues, as is normal practice for sanitary control.
- It is not suitable for rotations of wheat and maize or for wheat and sorghum because it is not possible to apply and incorporate herbicides against graminaceous weeds before sowing. This situation may change once there are selective post-emergence herbicides against these weeds for crops of maize and sorghum.
- It is not suitable when it is not possible to maintain a good residue cover over the soil.
- This system requires well-trained machinery operators.

Machinery

Machinery requirements are a rotary mower, direct seed drills both for small and large grains, a sprayer and a harvester. The seed drills for direct drilling, in order to work well, need to have a spring-loaded cutting disc located towards the front, which cuts through the residue cover and forms a slot or cut in the soil (Figure 24). The cutting disc can be flat, which makes it easier to cut through the crop residues, or it can be fluted or rippled, which gives better soil loosening along the narrow strip where the seeds will be placed. Fluted or rippled discs require a greater downward

FIGURE 24
Seed drill for direct drilling small grains

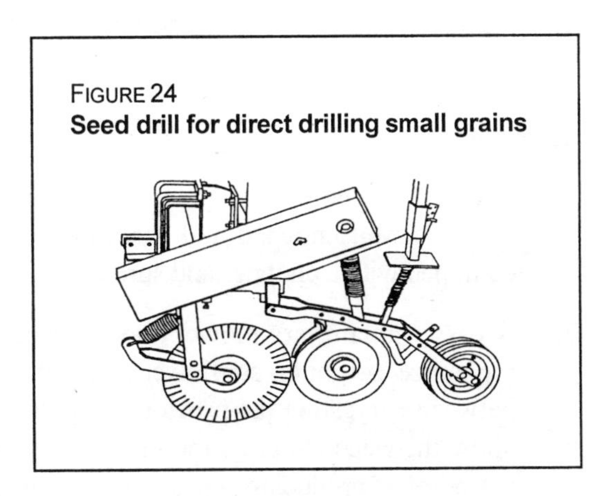

pressure in order to cut through residues and penetrate the soil and sometimes a disc that is flat on the external edge and rippled on the interior part can better accomplish the two functions.

Behind the cutting disc are located the double discs for opening the slit, which must have depth-adjustment wheels. Their purpose is to open a slot where the seed will fall. Sometimes, behind the double discs there is a disc or a knife to cover the slit with soil.

Towards the back a pressure wheel (or wheels) are located. There are many types of pressure wheel, simple, double and some with one or two rims. Their purpose is to place soil over the seed and to ensure a firm contact between the seed and the soil. The most appropriate type of pressure wheel depends on the soil texture and consistency and the amount of residue on the surface, and so it might be advisable to change pressure wheels from one field to another if the soil type changes.

Planters for maize, sunflower and cotton normally have additional hoppers for placing the fertilizer to one side and slightly deeper than the seed. The parts of the fertilizer unit are similar to those of the seeding unit. In contrast, there is less room to fit fertilizer hoppers in drills for small grains because the row spacing in these cases may only be 18 cm.

Pre-requisites

Before initiating a zero tillage programme it is important to determine whether the soil has any nutritional deficiencies, particularly phosphorous, which should be rectified before starting activities. In the same way, soils that are compacted should be loosened and any weed infestations eliminated. If there are serious weed problems, herbicides should be applied to the previous crop or alternatively, a cover crop should be sown to eliminate the weeds before starting zero tillage practices. In addition, for land with an uneven micro-topography, it is advisable to loosen the surface layer of the soil with a pass using a chisel plough followed by levelling the plot with a levelling harrow coupled to a spiked-tooth harrow. Although zero tillage is not very suitable for soils with drainage problems, if it is planned to practise zero tillage in these types of soil then drainage channels should be installed. It will also be necessary to install windbreaks in zones subject to strong winds.

It is better to commence zero tillage practices when there is a soil cover of at least 80 percent, for example with a crop that produces a lot of residues or with a cover crop. For the first two crops it is recommended to sow varieties that give large quantities of green matter and that allow good weed control. Crops of soybean and sunflower allow good control of both broad-leafed and graminaceous weeds, but only sunflower produces large quantities of green matter and hence a good residue production.

Operations

1. The first step before commencing zero tillage is to ensure that the residues from the previous crop are well chopped and spread evenly over the plot. For this, the harvester should be equipped with a chopper and spreader.

2. A pass with a rotary mower may be necessary in order to reduce the size of the stalks from the maize, sorghum, sunflower and cotton which remain after the harvest. During the period between harvesting the previous crop and sowing the zero tillage crop, it is important not to allow the weeds to grow too much. Once these reach a height of 15 cm or when they are on the point of producing seeds, they should be mown again. If there is sufficient soil moisture

to allow sowing a cover crop during this period, this is far preferable to just allowing the weeds to grow.

3. It is recommended that the weeds be eliminated through the application of herbicides such as glyphosate. Several studies in Bolivia have shown that good weed control can be achieved with the application of 2 l/ha of glyphosate mixed with 0.5 l/ha of 2.4-D amine, and with the addition of 3 kg/ha of urea so as to increase the efficiency of the glyphosate. In order to achieve good control, it is important to apply systemic herbicides when there is moisture in the soil, there is sunshine and the weeds are not too big. If the weeds are not transpiring well or there is no sun, systemic herbicides do not function well. If the weed-control operation falls on a windy day, care must be taken that the herbicide does not drift from the field towards other crops.

4. A cover crop may be eliminated by a single pass of a roller-crusher (a type of roller that crushes the weeds combined with herbicide application). If there is a considerable volume of cover crop, then one should await about a week for the foliage to dry out and for the volume to be sufficiently reduced so as not to cause problems during seeding.

5. The operation of the seed drill should be checked before sowing commences.

   i. Check that the depth of penetration of the cutting disc is between one and three centimetres more than the planned sowing depth. If not, then increase the spring pressure.

   ii. Check that the seed rate and the depth of seeding are suitable. If not, adjust the depth of the double disc furrow opener and the seed feeding mechanism.

   iii. Check that the soil moisture is sufficient for the slit to be closed and the seed to be covered with soil. If the slit does not seal, it is probable that the soil moisture content is too high for sowing. In this case, one should wait for a few days until the soil has dried out.

6. The rate of sowing will be about 70 percent slower than that achieved in a conventional system. Regularly check the sowing depth and the seed rate.

7. Control the weeds using an application of herbicides as necessary and wherever possible use integrated pest management techniques through the application of selective and biological insecticides.

8. Make sure that the harvester is well adjusted to chop the straw and to spread it uniformly over the surface of the field.

COMBINED TILLAGE AND SEEDING SYSTEMS

Combined tillage and seeding systems refer to systems where the soil preparation and seeding are done in a single pass. This requires special machinery made up of a number of components and for which there are many variations. The machinery tends to be very long due to the space needed for the different components, and to leave sufficient room for the movement of the soil and residues without causing problems of blockages. The three most common combined systems are strip tillage-seeding, ridge tillage-seeding and deep tillage-seeding ("rip-plant").

Strip tillage-seeding

In this system, a strip with a width of between 5 and 20 cm and to a depth of between 5 and 10 cm is prepared and drilled; the intervening soil between the strips remains undisturbed. There are variations in the type of machinery but most of them have a cutting disc in the front, followed by a tine or a disc to loosen the soil and behind this, a seeding unit similar to those in direct drills. Sometimes there are heavy press wheels located over the chisel point to prevent the formation of large aggregates.

The advantages of this system are that crop establishment and initial growth rates are both rapid because the soil is loosened along the tilled strips. The drill works well because the seed can be placed more evenly and can be better covered. In addition, it is easy to place the fertilizer in the loosened strip. The soil remains undisturbed between the strips, which results in better infiltration. In this system, less fuel and power is used than in conventional systems.

The system is suitable for hardsetting soils and for soils susceptible to compaction.

The main limitation of this system is that often, suitable machinery is not available and crusts may form along the strips.

Ridge tillage-seeding

The soil and the residues lying on the crest of the ridges shaped during the previous season are disturbed during a single pass and seeds are planted into clean, flat and smooth rows of the ridges. No tillage takes place between the rows before the sowing operation and they remain with a protective residue cover. Once or twice during the growth cycle of the crop, the weeds are controlled and at the same time, the ridges are reformed with a cultivator. This system corresponds to a type of controlled traffic because the wheels of the equipment stay in the same furrows and so do not compact the soil between the ridges. This system requires a planter which is equipped to remove the top of the ridges before sowing in the ridge. The system uses less herbicide, gives better crop establishment in the loosened soil of the ridges and is more suitable for poorly drained soils. The major limitation is that it requires the use of specialist machinery.

Deep tillage-seeding (Rip-plant)

This system is similar to zero tillage except that the seed drill has a subsoiler mounted between the cutting disc and the double disc furrow-opener (see Figure 25). In addition, pressure wheels have to be mounted behind the double furrow-opener so as to close the seeding slit. This system has all the advantages of zero tillage and has been specially developed for hardsetting and compacted soils. Limitations are the availability of suitable machines and the high draught force needed.

SUBSOILING

Subsoiling should be considered as a practice for recuperating soils that have been degraded due to serious problems of compaction. Generally speaking, subsoiling is not a tillage operation that should be used routinely every year for soil preparation.

Subsoiling has the effect of lifting, breaking and loosening the soil. This results in better root development and often in better soil drainage.

Advantages

- The main advantage is that it breaks up the compacted layers and loosens the soil without inverting it as occurs during ploughing. In this manner, the sub-surface soil is not brought up to the surface and the majority of the residues remain on top of the soil surface.

FIGURE 25
Equipment for deep tillage and seeding combined in one operation

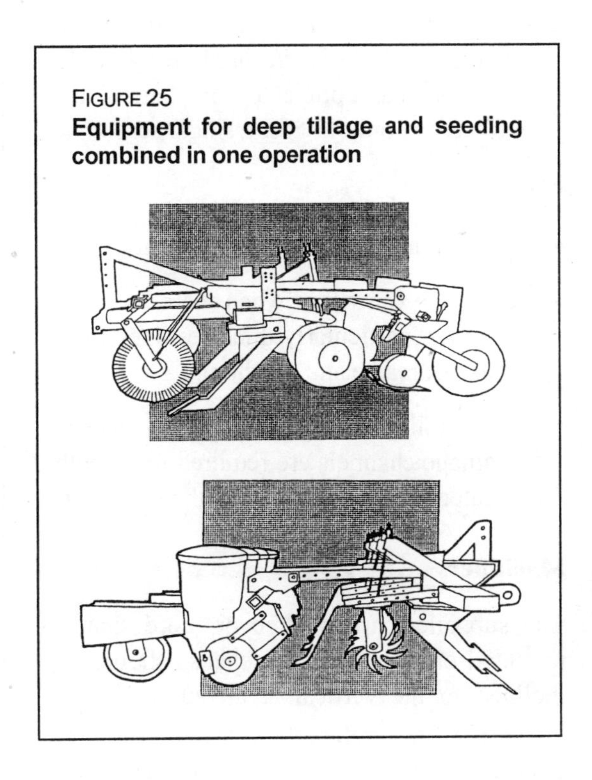

- The greater root penetration in well-drained soils can increase yields, particularly in areas suffering from a moisture deficit. One may also expect better yields in soils with drainage problems. Subsoiling clay soils often benefits both the crop roots and drainage of the soil and thus may overcome both moisture deficit problems during the dry season and problems of excessive moisture during the rainy season. Subsoiling gave yield increases of between 0.94 and 1.57 t/ha for maize and from 0.19 to 0.25 t/ha for soybean in North Carolina, USA compared to yields from the conventional system using a disc plough (Naderman, 1990). In Santa Cruz, Bolivia, subsoiling gave increases in the production of soybean for a highly compacted soil of between zero and 90 percent, depending on the seasonal rainfall. It was estimated that the minimum response to subsoiling in winter for seven years in ten would be 56 percent, which is equivalent to a gross profit margin of US $ 98/ha/year.

Limitations

- Subsoiling should only be carried out when the soil is dry or slightly moist, which is more difficult in clay soils. Subsoiling when the soil is dry requires a lot of power and often it leaves large aggregates with big spaces between them, providing unfavourable conditions for germination and initial plant growth. Subsoiling clay soils in moist conditions creates a channel where the point of the subsoiler has passed, without loosening the profile and breaking up the compacted layer (see Figure 26 showing the difference in the degree of soil loosening with a subsoiler under moist and dry conditions).

- Subsoiling leaves the soil very loose where the cuts or slits are made, which can hinder the establishment of the crops.

- Subsoiling when the soil is dry sometimes leaves very large aggregates on the surface, which obliges secondary tillage to be undertaken so as to create desirable seedbed conditions. These operations can cause compaction if it rains between the time of subsoiling and carrying out the secondary tillage operations.

- Subsoiling needs a lot of power and takes considerable time.

- The beneficial effect of subsoiling lasts only for a very short time in some soil types, particularly in hardsetting soils. For soils susceptible to compaction, the effect may only last for a single season.

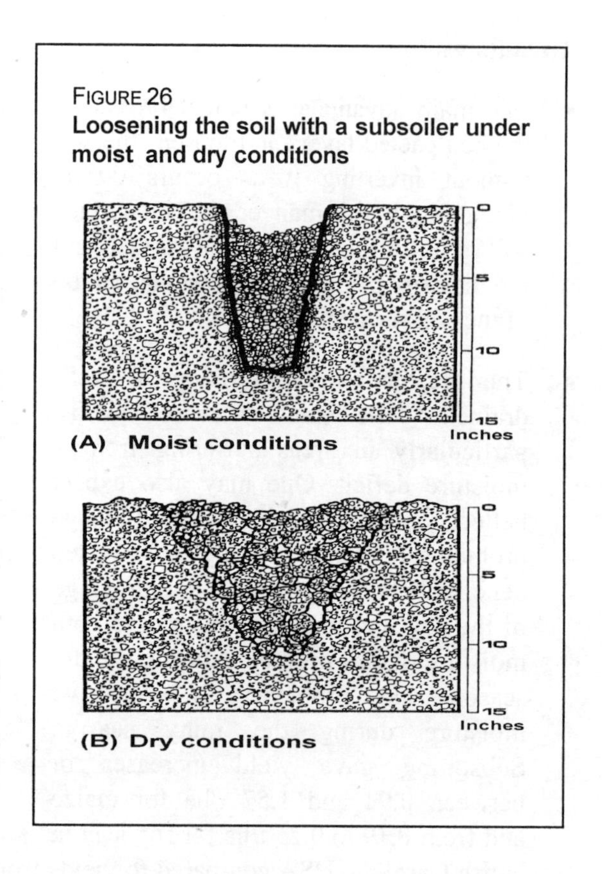

FIGURE 26
Loosening the soil with a subsoiler under moist and dry conditions

Pre-requisites

- The soil must be dry or slightly moist.
- Clogging can occur when there are crop residues, particularly thick residues from stalks of maize and sorghum.
- For soils with drainage problems, drainage channels are required to a depth greater than that reached by the subsoiler.

Machinery

The subsoiler carries three or more shanks mounted on a toolbar. The shanks should be inclined to the vertical at an angle greater than 25-30°, preferably 45°, and it is advisable that the height be adjustable (see Figure 27 for examples of different types of shank). The points of the shanks are normally about 1.5 inches wide and should be easy to replace. The condition of the point is very important and often the subsoiler fails to give good results due to the poor condition of its points.

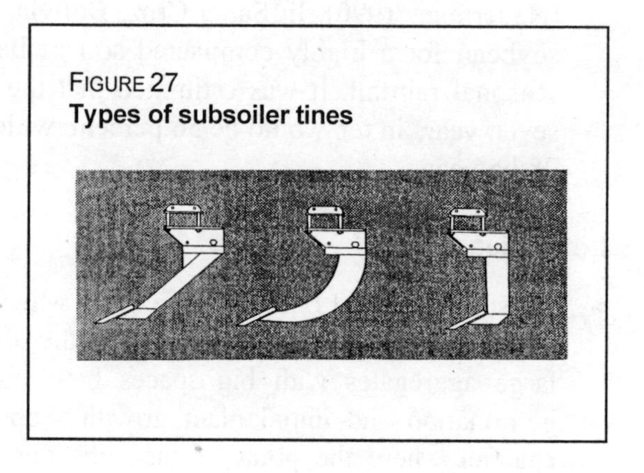

FIGURE 27
Types of subsoiler tines

Coupling other discs or rolls to the subsoiler can be useful. A cutting disc in front of the subsoiler shank eases operations in conservation tillage systems, a crumbler roll coupled behind the shanks will help to break up large aggregates and the addition of press wheels or discs helps to close the slits.

For combined systems of tillage and seeding, the subsoiler can be coupled to a planter ("ripper-planter") on to a disc ridger and planter ("ripper-bedder"). These systems have the advantage of preparing the land and sowing in a single pass.

Operation

The number of shanks and the spacing between them depends on the power of the tractor and the depth of work required. When the subsoiler shank passes through the soil, it loosens a triangular section of the soil (Figure 28). The width of the soil loosened at the surface approximates to the working depth of the shank. In order to ensure that the compacted layer is well loosened, the working depth of the shanks should be approximately equal to 1.5 times the depth of the bottom of the compacted layer. So as to ensure that there is a good overlap of loosening in the upper and lower parts of the profile, the tine spacing should not be greater than the working depth.

The power required for each tine depends on the state of compaction of the soil, the design of the subsoiler and particularly, on the state of the point. In general terms, the power requirement is between 20 and 30 HP per shank.

For a 90 HP tractor, where the bottom of the compacted layer is found at about 26 cm, the subsoiler should penetrate to a depth of 39 cm.

A 90 HP tractor can pull a subsoiler with three shanks and these should be adjusted so that one lies behind each tractor wheel and the third is located in the centre. The distance between tractor wheels is approximately 1.5 m and the tine spacing will thus be about 75 cm. With this spacing, complete soil loosening of the compacted layer will not be achieved, particularly halfway between the tines. Under these circumstances it would be advisable to make two intercalated passes of the subsoiler in such a way that the distance between the slits of the first and second passes is about 37 cm. Alternatively, one could use four shanks with a 100 HP tractor to a depth of 50 cm so as to more completely loosen the compacted layer. If a machine to subsoil and sow in a single pass is available, or if it is feasible to sow the crop in such a way that the rows coincide with the subsoiled slots, it will only be necessary for the subsoiler to work down to the bottom of the compacted layer, which would need less power (Figure 29).

For soils with problems of drainage, subsoiling should be undertaken in a

FIGURE 28
Representation of the relationship between the working depth of the subsoiler, the width of soil disturbance and the shank spacing

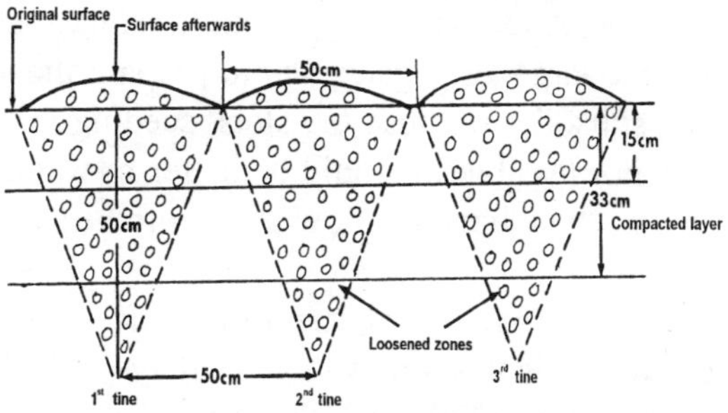

Original surface
Surface afterwards
50cm
15cm
33cm
Compacted layer
50cm
Loosened zones
50cm
1st tine
2nd tine
3rd tine

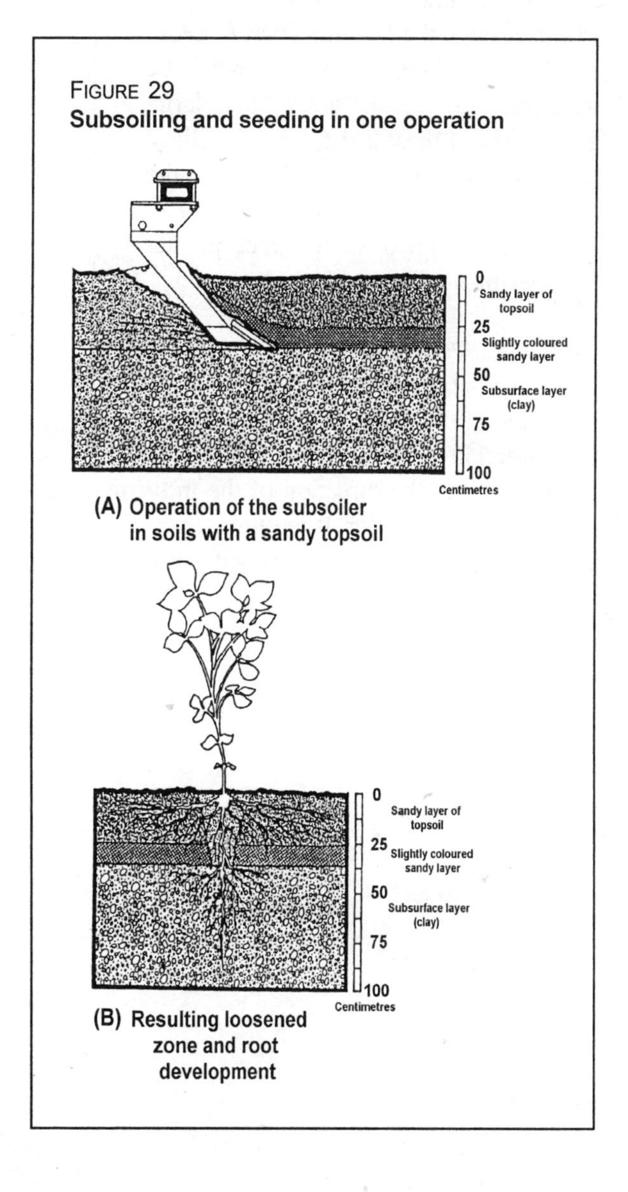

FIGURE 29
Subsoiling and seeding in one operation

direction perpendicular to the line of the drainage canals so as to ease the flow of water towards the drains. In conventional tillage systems, both subsoiling and primary tillage operations should be carried out during the dry season after the harvest and before preparing the seedbed. In the event that subsoiling is done after soil preparation, the tractor wheels might not grip and there will be worse problems of slipping, causing a serious risk of soil compaction. A rotary mower should be used before carrying out a subsoiling operation to avoid problems of residue blockages. After subsoiling, it might be necessary to carry out further tillage operations to break down the large aggregates and to prepare the seedbed. If discs and press wheels can be hitched to the subsoiler to close the slots, seeding may be undertaken as soon as the rains arrive but if this is not feasible, it would be better to wait some time for the loosened soil to settle. Farmers often say "you have to allow the soil to heal itself before sowing".

Rotary mowing of the residue is necessary in conservation tillage systems, particularly in the case of maize and sorghum and it is advisable that the subsoiler is equipped with discs to cut through the residues so as to avoid blockages. Afterwards it will be necessary to break down the large aggregates, if these exist, by strip tillage prior to seeding. In this case, the combination of the subsoiler with discs and rolls, coupled to a direct planter ("ripper-planter") would be appropriate.

FIGURE 30
Operating principle and construction of a "Paraplow" (Source: Wieneke, F. and Friedrich, Th. 1989)

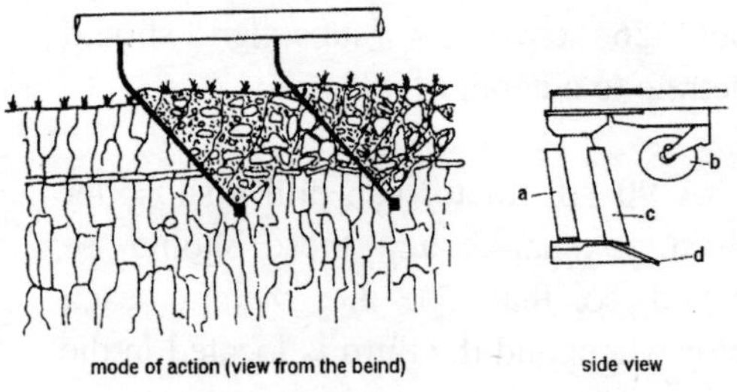

The "Paraplow" is similar to a subsoiler but the lower part of its shank is inclined in two directions, forwards in the direction of movement and also at right angles to the direction of the movement. The "paraplow" has various advantages over the conventional subsoiler. It requires less power and does not bring aggregates from the subsoil up to the surface, as is often the case with subsoilers (Figure 30).

Chapter 9

Land use according to its capability

Many farmers throughout the world are using their land well and in a sustainable manner, based on generations of experience. Others are doing the same on the basis of extension advice or experimental results. But in many places, new land had been cleared for agriculture, or the land use has changed drastically because of population changes or economic pressures. Often, such changes in land use have been made without any previous studies to show what would be the most appropriate use and what effect on the environment the different uses will make – and in a manner and in places that are inappropriate. This has resulted in poverty, environmental degradation, inefficient economic exploitation and loss of natural resources such as soil and water. The best uses for land depend on economic, social, political and cultural conditions, as well as on the soil characteristics and their response to the use.

In an economic sense, land has many attributes such as farm or parcel size, proximity of water supplies, transport facilities and markets (FAO, 1967). According to FAO (1993), land is a segment of the earth's surface defined in space and in function of its characteristics and properties including the attributes of the biosphere which are reasonably stable or cyclically foreseeable. These include those of the atmosphere, the soil, the geological substratum, the hydrology and the result of present and past human activities insofar as these attributes have a significant influence on the present and future use of the land by man.

LAND EVALUATION

Land varies in its characteristics from place to place. This variation affects its use as for each type of land, there are uses that are more or less physically and economically appropriate as regards productivity and return on invested capital. The variation in land is largely systematic and is caused by known factors which means that they can be mapped, separating them into homogenous areas. These particular areas have a behaviour that is foreseeable with some degree of certainty when submitted to certain types of use. The degree of certainty depends on the quality of the available information and on the knowledge used to relate these land characteristics to its response to use.

Land evaluation is the process of evaluating the response of the land when it is used for specific purposes. This process allows rational land use planning to be undertaken through the identification, for each type of land, of uses that are adequate and environmentally and

E. Giasson
Soils Department of the Federal University of Rio Grande do Sul
Porto Alegre, Brazil

economically sustainable for the natural and human resources. In this manner, it can be an important tool for land use both for individual users, for groups or for society as a whole.

A system for evaluating the aptitude of land use should be based on objective parameters which can be applied on whatever scale from initial reconnaissance to individual planning for properties. It should be adequate for local conditions and should consider economic aspects involved in each type of use so that it is applicable in the majority of situations concerning the availability of natural resources.

For technical classification, the individual cases are grouped together in accordance with the main characteristics of practical and specific interest. These concern agricultural performance of the soils, physical and socioeconomic aspects. The classification is thus interdisciplinary in nature. This type of classification is a process of estimation concerning the performance or the aptitude of the land use when it is used for specific purposes (Resende *et al.*, 1995).

Two main systems are used in Brazil for soil survey work aimed at classification of the suitability of the soils:

a. the System of Classification of Land Use Capability (Klingebiel and Montgomery, 1961);

b. the System of Evaluation of Land for its Suitability for Agricultural Use (Bennema *et al.*, 1964, modified by Beek, 1975).

In addition, other systems are also used with similar aims, having been developed to be better adapted to specific conditions, such as the system presently being used by the Federal University of Río Grande do Sul.

A classification system for the suitability of the soils must respond to a series of questions such as suggested by Brinkman and Smyth (*apud* Klamt, 1978):

a. How is the soil being used and what will happen if its present use is not altered?

b. What other uses for the soil are possible under existing social and economic conditions?

c. What alternative use or uses show possibilities for maintaining the quality of the environment?

d. What limitations or adverse effects are associated with each type of use?

e. What investment is necessary to minimize the limitations and adverse effects?

f. What are the benefits for each alternative use?

g. Is there provision for major modifications to the soil use or the management system? What are these? How will they be realized? What is the investment? What are the benefits? Who will benefit?

System for the classification of land use capability

Klingebiel and Montgomery (1961) of the United States Soil Conservation Service set up this system, and it was adapted and introduced to Brazil by Marques (1971). Afterwards, various other approximations were introduced such as those done by Lepsch (1983) and Lepsch (1991).

The system is recommended for use in planning soil conservation practices either for private farms or agricultural enterprises, or for small hydrographic catchment areas. It should only be used for other objectives, such as regional studies, after modifications and accompanying studies that consider the socioeconomic conditions and the agro-climatological suitability for the crops (Lepsch, 1991).

The system is based on interpretation of the characteristics and intrinsic properties of soil and environment. It is adapted to a high technological level of the farmers. The objective of the system is to identify homogeneous classes of land and to define their maximum utilization capability without incurring risks of soil degradation, particularly accelerated erosion. The system takes into account the permanent limitations of the land that are related to the possibilities or limitations of its use. However, it also considers both socioeconomic and agricultural policy issues (Lepsch, 1991).

The hierarchical structure of the system is as follows:

- *groups concerning suitability of utilization* (A, B and C): these are established on the basis of the types of intensity of land use;
- *classes concerning suitability of utilization* (I to VIII): these are based on the degree of limitation for use;
- *sub-classes concerning suitability of utilization* (IIe, IIIe, IIIa, etc.): these are based on the nature of the limited use;
- *suitability of utilization units* (IIe-1, IIe-2, IIIa-1, etc.): these are based on specific conditions that affect land use and management.

There are thus a total of eight classes organized into three orders, with the intensity of land use decreasing from I to VIII. Group A includes arable land, suitable for annual and perennial crops, pastures, reforestation and agro-pastoral land, including the land suitability classes I, II, II and IV. Group B comprises land adapted for pasture, forestry and agroforestry, including the land suitability classes V, VI and VII. Group C consists of non-arable land but which is suitable for the protection of forestry flora and fauna, together with water storage. It comprises class VIII land.

Classes II to VII, except for class V, are subdivided into sub-classes according to their permanent limitations related to erosion (e), soil (s), water (a) and climate (c). The suitability units describe more explicitly the nature of the limitations, so they facilitate the process of establishment of land management practices.

The land capability classification system has the advantage of the fact that it can be used for planning purposes and can indicate conservation use and appropriate practices at farm level. A disadvantage is that the system presupposes a use incorporating an advanced level of soil management practices and it does not consider the management levels developed, for instance, by farmers using draught animals and so risks to under-estimate the agricultural potential of certain farms. Following this approach, stony land with slight slopes, which can be successfully cultivated under annual crops with animal traction, would be evaluated as class VI land and denominated as "inappropriate for annual crops" by the system as they cannot be cultivated by mechanized tractor-operated equipment.

Furthermore, the system needs a basic soil map of the area or a detailed map of the physical attributes of the land, which are not always available. Further difficulties in estimation also arise from the conservation land utilization recommendations and the rigidity with which the erosion risks are interpreted and mapped, particularly for the class VI and VII lands (land inappropriate for use with annual crops). These difficulties are aggravated by the lack of scientific information regarding recommendations for the use of these areas (Streck, 1992).

Other limitations of the system, according to Klamt (1978), concern the absence of information regarding certain parameters used for the definition of the units for utilization capability, such as production data for the main crops and an exact definition of the management conditions under which these production figures were obtained. Furthermore, the problems related to soil erosion are highlighted, without detailed consideration of the limitations at the level of natural soil fertility which constitutes a grave omission for regions under development conditions.

System for the evaluation of land for its suitability for agricultural use

This system was first proposed by Bennema *et al.* (1964), reformulated by Beek (1975), by Ramalho Filho *et al.* (1977), Ramalho Filho *et al.* (1978) and by Ramalho Filho and Beek (1995). The system is also known as the "FAO/Brazil System".

The system is structured in the following categories:

- *management levels* (A, B and C): these diagnose the performance of land at different technological levels;
- *groups of agricultural suitability* (1 to 6): these identify in the map, the most intense feasible type of land utilization;
- *sub-groups of agricultural suitability* (1ABC, 1bC, etc.): these indicate the type of land use according to the level of management;
- *classes of agricultural suitability* (good, normal, limited and unsuitable): these express the agricultural suitability of the land for a particular type of use, according to a defined level of management and within a particular suitability sub-group.

The management level A (primitive) corresponds to the lowest technical and cultural level, characterized by the absence of any modern technology, the use of manual labour or animal traction and simple implements and tools. Management level B (little developed) represents an intermediary stage with low capital investment for the land improvement and conservation and with mechanization based on animal traction or tractors used exclusively for brush clearance and initial tillage operations. Management level C (developed) is based on the highest level of technology. It is characterized by high capital investment and the application of research results concerning land management, improvement and conservation, together with the use of motorized mechanization for most phases of the agricultural operations (Ramalho Filho and Beek, 1995; Resende *et al.*, 1995).

Six land suitability groups are situated within the highest level of classification, being essentially comparable to the eight classes of use in the North American system. Groups 1, 2 and 3, apart from identifying cropping as a type of land use, have the objective of representing within the sub-group, the main suitability classes for the land recommended for the crops according to different management levels. Groups 4, 5 and 6 only identify types of use which are, respectively, sown pasture, agroforestry and/or natural pasture, and preservation of flora and fauna. This identification is done independently from the suitability class (Ramalho Filho and Beek, 1995).

The agricultural suitability sub-group is the result of the joint evaluation of the suitability class as related to the management level. There are differences as regards the second level of classification, when the system is compared with the North American one. The sub-group refers to agricultural suitability of the land for the types of use adopted, whereas in the North American system reference is made to the types of limitation that determine each class. This means that in the example of land being described as "1(a)bC", the number 1 indicates the group with the highest suitability class of the components within the sub-groups. Within the management level C (group 1), the land is classed as of good suitability, under management at level B (group 2), the land is of average suitability and with level A management (group 3), the land has only restricted suitability. The absence of any of the letters signifies that it is not suitable for any use (Ramalho Filho and Beek, 1995).

The link-up of the classes in this system is done through a parametric process and the parameters that define each class are set out in guide-tables. These define the different types of soil use and consider the following limiting factors: deficiency in fertility, deficiency in water, excess of water or deficiency in oxygen, susceptibility to erosion and mechanization impediments. The following degrees of limitation are defined for each factor: nil, slight, moderate, strong and very strong (Streck, 1992). These degrees of limitation refer to the relative divergences of the soil as compared to a hypothetical "ideal" soil (Resende *et al.*, 1995).

The guide-tables for evaluation of the land agricultural suitability constitute a general orientation for agricultural suitability classification. This is determined by the land's degree of limitation as related to the limiting factors for the three management levels A, B and C. In this manner, the class of the land agricultural suitability is obtained. It reflects the strongest degree of limitation and is in accordance with the established management level. These guide-tables should only be used as a general guide as the evaluation will vary according to the local peculiarities, the quality and diversity of the data and also according to the detail of the study (Ramalho Filho and Beek, 1995).

The system can also indicate, for management levels B or C, the feasibility for improving the current agricultural condition of the land. This is expressed in underlined numbers. These accompany the letters that represent the stipulated degrees of limitation in the guide-tables.

The system has the advantage of identifying the land suitability for each of the management levels under consideration. It is recommended for areas that have pedological surveys to either an exploratory or reconnaissance level and where regional agricultural planning and agricultural zoning are required. A disadvantage is that it does not specify appropriate management practices at farm level and it is based on pre-established guide-tables that are applicable to the generalized regions of the whole country and which are not always applicable to local conditions. Due to this, the system generalized and over-estimated the agricultural potential of areas in the basalt shield of Río Grande do Sul in Brazil (Streck, 1992).

When this evaluation system is used at the level of individual farms or in small catchment areas, it needs to be adapted to the local conditions. These adaptations should consider management levels that are less well developed than those in the North American system and should be more specific than those in the FAO/Brazil system as regards recommendations for conservation practices at farm level.

Parametric method for classification of land use capability

According to Streck (1992), this method classifies the land on the basis of characteristics of the soil and of the physical environment, which represent the different degrees of limitation for agricultural use of the land and which serve as parameters to differentiate between farms. This system combines the favourable aspects of the North American and FAO/Brazil systems. It considers less developed management levels and indicates specific conservation practices so different farms can be exploited without exceeding tolerable limits of soil losses. The method should be developed to be simple so that it may be easily used by different technicians and so that consistent classifications and management recommendations are obtained. The method is based on the establishment of guide-tables that have been established as a result of research studies, the experience of the evaluator and empirical observations of the farmer. They allow definition of the most appropriate conservation practices for the different farms and the type of land use. This method has the advantage of being adequate for the evaluation of the potential use of soils at farm level and within hydrographic catchment areas. It accommodates less developed management systems and is easy to apply and to understand by technicians. It defines commonly occurring groups of soils and which present different degrees of limitations, developed into specific guide-tables that are simpler for each group.

A typical guide-table is illustrated in Table 18. It has been developed for the determination of the classes of agricultural suitability for the soils in the catchment basin of Lageado Atafona, in Santo Ângelo (Brazil), where three groups of land were defined. The table refers to land under group 1, characterized by deep soils, well drained, stone-free and with limitations posed according to different classes of slope, depth and erosion (Streck, 1992).

TABLE 18
Guide-table for definition of the classes and sub-classes of land use capability for group 1 soils of the Lageado Atafona catchment basin, Santo Ângelo, Brazil (Streck, 1992)

Slope classes	Degree of erosion	Soil depth		
		Very deep (H1) > 200cm	Deep (H2) 100-200cm	Shallow (H3) 60-100cm
(1) 0 - 5 %	-	C1	C1	C2
(2) 5 – 10 %	Pe	C2	C2	C3
	y	C5	C5	C5
(3) 10 – 15 %	Pe	C3	C3	C4
	y	C5	C5	P7
(4) 15 – 20 %	Pe	P7	P7	P7
	y	R9	R9	R9

Pe – little erosion; y - eroded; C – annual crops; P - pastures; R - reforestation; 1,..., 9 – conservation practices (sub-classes).

A study was undertaken in the basalt shield of Rio Grande do Sul to define the suitability for the soils to be used through adequate management practices. Guide-tables were proposed, based on the land slope, the soil depth and the amount of stones present. Limitations regarding fertility, drainage and others were omitted to simplify the tables and to make interpretation easier (Klamt and Stammel, 1984). The authors of the report support the idea that, in order to avoid considerable losses through erosion and nutrient shortages, and so as not to cause environmental imbalances, the use of the soils in this region should be based on their suitability or capability.

In a study similar to that described above, guide-tables were also developed for the appropriate use of soils in the West and in the valley of Río do Peixe, in Santa Catarina (Brazil), where 75 percent of the area is hilly and stony land (Uberti, 1985).

Chapter 10

Soil cover

Since the beginning of the twentieth century, various studies have been undertaken to observe the effects of a vegetative cover on the reduction of runoff and erosion. The first observations recognized the effect of the cover in the protection and maintenance of the soil pores and the consequent reduction in surface runoff. However, it was not until the early 1940s that a quantitative evaluation of the effect of vegetative cover was made. Studies made in Ohio, USA with a rainfall simulator (Borst and Woodburn, 1942), showed that by intercepting the impact of the raindrops with a straw cover, erosion was reduced by about 95 percent.

Kohnke and Bertrand (1959) reported that a soil cover of between two thirds and three quarters of the surface would be sufficient to protect the soil from the impact of the rain and practically eliminate the transport of soil material by the droplets. This cover corresponds to approximately five tons per hectare (t/ha) of straw.

Mannering and Meyer (1963) verified that 2.5 t/ha of wheat straw were sufficient to offer an 87 percent coverage of the soil and to completely eliminate runoff and erosion.

Meyer *et al.*(1970) observed that the soil physical conditions (texture, permeability) and the land slope had an important influence on the efficiency of the soil cover.

a. 4.48 t/ha of straw were sufficient to contain the erosion of a loam textured soil with low permeability, when not cultivated and on a 15 percent slope;

b. 2.24 t/ha of straw were sufficient for a silt-loam soil when cultivated, with moderate permeability and on a three percent slope;

c. 1.12 t/ha of straw were sufficient for a silt-loam soil when cultivated, very permeable and with a five percent slope.

Lal *et al.* (1980) concluded that crop residues could prevent erosion and sustain production in soils that are difficult to manage. The magnitude of these effects depends on the quality of the residues (the cover), and also on improvements to the physical and chemical conditions of the soil. However, the authors noted that the results could differ from temperate to tropical regions.

L. do Prado Wildner
Enterprise for Agricultural Research and Rural Extension of Santa Catarina (EPAGRI) - Small Farm Research Centre (CPPP)
Santa Catarina, Brazil

Soil cover effects on soil and water losses

The soil cover provides a protective action by intercepting and absorbing the direct impact of the raindrops, thus preventing the surface becoming sealed and preserving the soil structure immediately underneath (Adams, 1966). In this manner, the infiltration of the water can be maintained throughout the rainstorm (Musgrave and Nichols, 1942). It follows that by increasing the soil cover, there is a reduction in the break-up and movement of the soil due to the rainfall droplets (Singer *et al.,* 1981). There is also a reduction in the runoff velocity and of the transport capacity of this surface flow (Lattanzi *et al.*, 1974; Meyer *et al.*, 1970; Mannering and Meyer, 1963). According to Singer and Blackard (1978), the quality and quantity of the residues affect the volume of the surface runoff. This is due to them delaying the onset of runoff, delaying the time it takes for the runoff to build up, and also by reducing the time between the end of the rainstorm and the moment of termination of the runoff (Table 19).

TABLE 19
Evaluation of the effect of increasing the quantity of maize residues in the soil cover on the runoff flow, the runoff velocity and the total soil losses. UFRGS, Guaíba, RS, 1984.

Quantity of residues (t/ha)	Soil cover (%)	Runoff flow			Velocity after 30 min. (cm/s)	Velocity after 60 min. (cm/s)	Soil loss (t/ha)
		Start (min.)	End (min.)	Total (min.)			
0	0	15	67	68.5	24.0	24.0	38.1
0.5	14	15	67	62.3	17.0	20.0	20.0
1	19	14	76	68.7	16.9	20.0	31.1
2	38	15	77	57.0	12.0	17.0	22.0
4	69	17	80	46.5	8.6	10.0	5.6
6	82	17	80	39.3	5.4	7.5	1.8
8	90	20	80	50.1	5.0	4.4	1.6
10	96	20	80	44.1	3.4	4.4	0.7

Soil: San Jerónimo (Red-yellow podsolic soil with sintering), average slope 7.5 percent.

According to observations made by Meyer *et al.* (1970), if harvest residues are left so that they are placed on the surface along a contour, this encourages the formation of a miniature staircase due to the accumulation of soil above the pieces of straw. Lattanzi *et al.* (1974) described the formation of small bunds that held back the surface runoff, increasing the depth of the film of water on the soil and so allowing part of the energy of the rain droplets to be dissipated.

Soil cover as affected by plants and their residues

Soil cover is the single most important factor in the control of erosion by water (Amado, 1985). The soil cover can be thought of as being provided basically by the cover afforded by the vegetative cover of the plants as they develop (during the vegetative period) or by their residues (Forster, 1981, cited by Lopes, 1984).

The soil cover provided by plants during their growth cycle varies from species to species in accordance with the phenologic and vegetative characteristics of each (cycle, growth characteristics, height, speed to cover the soil, stage of growth). It also depends on the cultural practices needed for establishing the crop such as the plant density, seed spacing, fertilization and liming (Table 20). Annual crops, for example, allow the soil to remain unprotected mainly during the period of soil preparation and sowing, and until the crop becomes completely established. The behaviour of perennial crops is totally different to that of annual crops and also differs from one perennial crop to another (Table 21).

The results of research undertaken in Paraná, Brazil, (IAPAR, 1978), showed that the total soil losses follow the increasing order of susceptibility to erosion: maize < wheat < soybean < cotton (Table 22). The data show that the total soil losses in the maize crop reached only 1.7 percent of the loss compared to bare soil without vegetative cover. The results also highlight the importance of protecting the soil during the initial period of the crops. It is during this period that practically the entire erosion problem is concentrated. Even then, the maize crop presents the smallest soil losses when compared to soybean, wheat and cotton. After the initial period of crop establishment, some 30 to 40 days after emergence, there is an effective protection of the soil by the crops and a consequent considerable reduction in the soil losses.

Lopes (1984) argued that despite the progressive reduction of erosion during the vegetative stages of crop growth, the crops do not reduce the erosion as efficiently as crop residues left in direct contact with the soil surface. For this reason, the use of crop residues as a soil cover is the most efficient, simple and economic way to control erosion (Amado, 1985).

TABLE 20

Effect of the annual crop type on soil losses by erosion. Average rainfall 1 300 mm and a slope between 8.5 and 12.8%

Crop type	Soil loss (t/ha)	Water loss (% of the rainfall)
Papaya	41.5	12.0
Beans	38.1	11.2
Cassava	33.9	11.4
Groundnut	26.7	9.2
Rice	25.1	11.2
Cotton	24.8	9.7
Soybean	20.1	6.9
Bocconia frutescens	18.4	6.6
Sugar cane	12.4	4.2
Maize	12.0	5.2
Maize + beans	10.1	4.6
Sweet potato	6.6	4.2

Source: Bertoni *et al.*, 1972.

TABLE 21

Effect of the type of perennial crop or vegetation on erosion losses of soil. Weighted averages for three soil types in the State of São Paulo, Brazil

Type of use	Soil loss (t/ha)	Water loss (% of the rainfall)
Forest	0.004	0.7
Pasture	0.4	0.7
Coffee	0.9	1.1

Source: Bertoni *et al.*, 1972.

TABLE 22

Losses of soil and water during the growth cycles of soybean, wheat, maize and cotton in a red dystrophic latosol soil and a slope of 8%. Londrina, IAPAR, PR. (1977)

Crop (established by conventional tillage)	Crop stage								Total	
	I		II		III		IV		Soil (kg/ha)	Soil (%)
	Soil	Water	Soil	Water	Soil	Water	Soil	Water		
Soybean	6 738	38.9	39	8.3	7	3.3	641	15.2	7 425	7.2
Wheat	2 216	52.8	1 755	50.4	6	2.3	691	25.8	4 668	4.5
Maize	994	17.8	747	8.5	35	2.4	0	0.0	1 776	1.7
Cotton	9 252	22.2	1 303	9.2	2 088	20.6	352	5.5	12 995	16.6
Bare soil	25 225	28.5	25 191	31.9	27 355	34.7	25 225	28.5	102 996	100

Adapted from IAPAR, 1978.
Crop stages: I = Germination to 30 days, II = 30 to 60 days, III = 60 days to flowering, IV = After harvest.

The effect of the crop residues on the control of erosion varies according to the quantity (Table 23), the quality (Table 24), the soil cover (Table 19), the management (Table 25) and the degree of decomposition of the residues (Cogo, cited by Lopes, 1984). It is clear from the tables mentioned above, that there are various interactions between the soil cover and the quantity of residues, between cover and quality, and cover and management. For this reason, for the same quantity of residues there are differences in the percentage of soil cover, depending on the crop type and the management given to the residues. For example, for the same quantity of organic matter (kg/ha), the residues of wheat offer a superior percentage of soil cover than the residues of maize and this, more than the percentage provided by soybean residues. On the other hand, a

50 percent soil cover with maize residues reduces the erosion by about 90 percent (Lopes, 1984).

TABLE 23
Total soil losses in plots with a 7.5% slope for a Red-yellow podsolic soil under simulated rainfall of 64 mm/h and with different quantities of crop residues

Quantity of residue (t/ha)	Soil losses (t/ha)		
	Maize residues	Wheat residues	Soybean residues
0	37.5	18.1	14.2
0.5	28.8	11.0	10.7
1	26.3	8.7	8.4
2	23.2	3.3	5.8
4	5.3	1.8	4.5
6	1.5	0.2	1.8
8	1.2	0.4	1.7
10	0.5	0.2	1.1

Source: Lopes, 1984

TABLE 24
Chemical composition of some residues used as a dead vegetative cover

Material	C/N Ratio	N (%)	P_2O_5 (%)	K_2O (%)
Black oats	36.25	1.65	0.21	1.92
Italian Ryegrass *	44.20	1.34	0.15	3.13
Hairy peas*	18.65	1.88	0.22	2.76
Common peas *	18.62	2.02	0.29	2.52
Serradella*	22.43	1.79	0.32	4.27
Chick pea*	18.79	2.23	0.22	3.49
Colonial grass	27.00	1.87	0.53	-
Elephant grass	69.35	0.62	0.11	-
Bermuda grass	31.00	1.62	0.67	-
Bahia grass	36.00	1.39	0.36	-
Coffee straw	31.00	1.65	0.18	1.89
Maize straw	112.00	0.48	0.35	1.64
Maize cob	72.72	0.66	0.25	-
Rice straw	53.24	0.77	0.34	-
Rice husk	39.00	0.78	0.58	0.49
Sawdust	865.00	0.06	0.01	0.01
Cassava stalks	67.14	0.70	0.25	-
Silkworm chrysalises	5.00	9.49	1.41	0.76
Silkworm debris	17.00	2.76	0.69	3.65
Cane bagasse	22.00	1.49	0.28	0.99

* Material in full flower.
Source: Calegari, 1989.

TABLE 25
Percentage of soil cover as a function of the management of residues from different crops

Type of residue	Implement used	Soil cover (%)
Maize	Scarifier	63
	Disc harrow	52
	Incorporation	17
Oats	Scarifier	77
	Disc harrow	72
	Incorporation	16

Source: Adapted from Sloneker and Moldenhauer (1977).

Chapter 11

Contour farming

Farming along the lines of equal contour is one of the most simple and efficient practices for the control of erosion. It consists of planting the crops according to the curved lines which follow the land surface at equal heights above sea level, or in other words, perpendicularly to the lines of steepest slope gradient (Sobral Filho *et al.*, 1980).

Contour farming requires the application of systematic tillage practices before the crop may be established. In this way, terracing and all soil preparation exercises (ploughing, scarifying, harrowing) must be carried out along the lines of the contours and because of this, the terraces will serve as a general guide for the direction of planting.

Contour farming is only recommended as an isolated erosion control measure for limited areas where the slope is less than three percent and the slope length is not long (Río Grande do Sul, 1985). In other conditions, the crop grown along contours must always be associated with other practices of conservation.

A study undertaken in the Campinas Institute of Agronomy (Bertoni *et al.*, 1972) showed a major increase in maize yields for the crop planted along the contours compared with down the slope (Figure 31). A smaller increment in yield was achieved by also preparing and tilling the land along the contours. When the soil preparation and the planting were both done down the slope, the furrows and the crop rows directed the flow of the runoff downwards, dragging the soil, the nutrients and the organic matter with it. When the crop is planted along the contours, this corrects the negative effect of the furrows left after preparing the land down the slope. By combining both the land preparation and the planting along the contours, small contour ridges are shaped. These, together with the planted crop, serve as obstacles, causing slight flooding and will thus increase the infiltration of the water into the soil and reduce the erosion (Table 26). This effect is very noticeable in conservation cropping where scarifiers are used, in situations where ploughing operations are reduced and even in areas where small farmers use animal traction ploughs.

TABLE 26
Effect of management and conservation practices on erosion losses under annual crops

Cultivation and planting practices	Soil loss (t/ha)	Water loss (% of rainfall)
Down slope	26.1	6.9
Along contours	13.2	4.7
Contours + alternation with pasture	9.8	4.8
Contours + bands of sugar cane	2.5	1.8

Source: Bertoni *et al.*, 1972.

L. do Prado Wildner
Enterprise for Agricultural Research and Rural Extension of Santa Catarina (EPAGRI) - Small Farm Research Centre (CPPP)
Santa Catarina, Brazil

FIGURE 31
Effect of the direction of planting and the soil preparation method on maize production
(after Bertoni *et al.*, 1972)

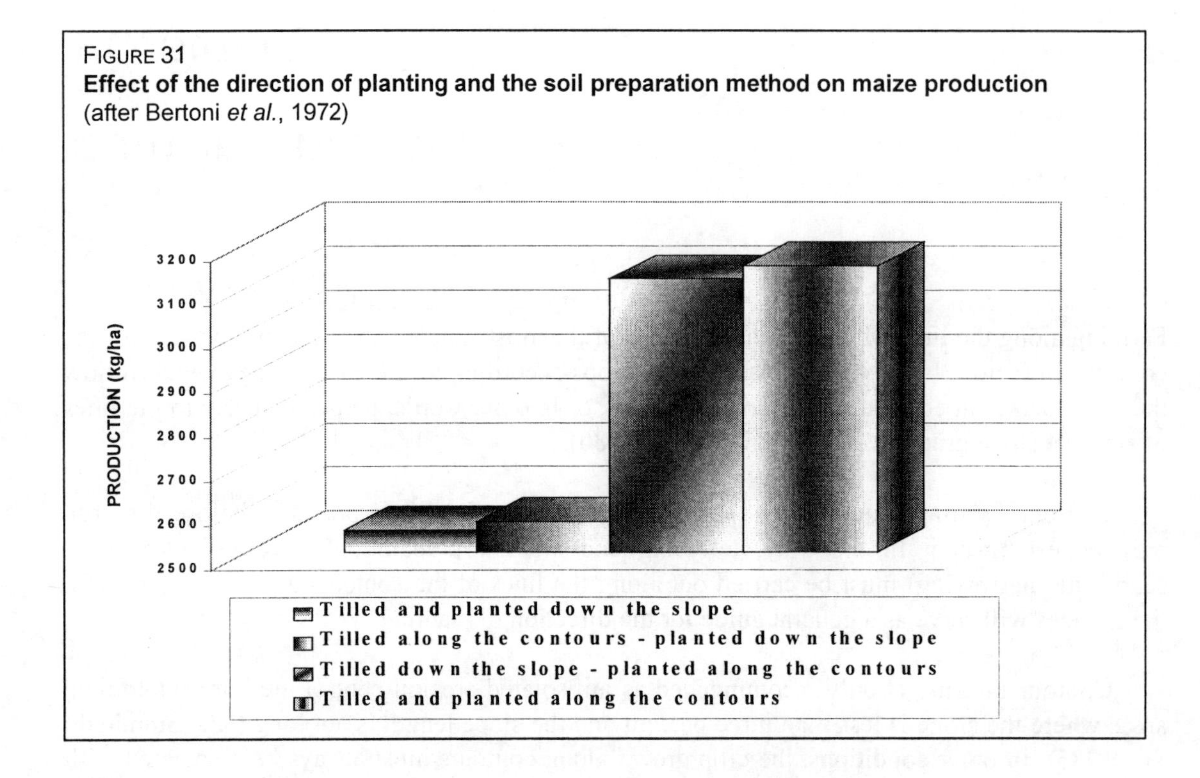

Chapter 12

Green manure

CONCEPT

The use of green manure has been recognized as an agricultural practice for millennia. It consists of incorporating non-decomposed vegetative matter into the soil with the objective of conserving or restoring the productivity of agricultural land. Traditionally, legume crops such as lupine were often used.

Green manure is considered today as a crop in rotation, following or associated with commercial crops, and incorporated into the soil or left on the surface as mulch to offer protection. The objective is to maintain, improve or restore the physical, chemical and biological properties of the soil (Costa *et al.*, 1992). Eventually, part of the green manure can be used for animal or human food, the production of fibres or the production of forage (Miyasaka, 1984). This is an important stimulus for adopting the practice as the various uses of green manure on the farm increase its benefits. With this new focus, plants other than legumes are more commonly used, including graminaceous, cruciferous and cariophilaceous crops.

FUNCTIONS OF GREEN MANURE

- Green manure protects the surface layer of the soil from high intensity rainstorms, sun and wind.
- It maintains high rates of infiltration through the combined effect of the root system and the vegetative cover. After decomposition, the roots leave pores in the soil and the cover prevents the break-up of aggregates and capping of the surface, and also reduces the runoff velocity.
- It provides a considerable and continuous supply of biomass to the soil, so maintaining and even raising during the course of several years, the organic matter content of the soil.
- If used as a mulch, reduces soil temperature variations and reduces evaporation, so increasing moisture availability to the commercial crops.

L. do Prado Wildner
Enterprise for Agricultural Research and Rural Extension of Santa Catarina (EPAGRI)
Small Farm Research Centre (CPPP)
Santa Catarina, Brazil

- The root systems break the compacted layers and cause aeration and structuring of the soil, or process that might be called *biological soil tillage*.
- It increases the recycling of nutrients. The root system, which is well developed in many types of green manure, has the ability to extract the nutrients from the deeper layers of the soil, so making them available for future crops once the green manure has decomposed.
- It reduces leaching of the nutrients. Heavy and intense rainstorms cause intense leaching of soluble nutrients from the soil. The green manure reduces this effect by holding the nutrients in the biomass and freeing them gradually during its decomposition.
- It adds nitrogen to the soil by the biological fixing process of the legumes. This can bring about significant economies in supply of nitrogenous fertilizer to the commercial crops, in addition to improving the soil nitrogen balance.
- It reduces the weed population through the suppressive (and in some cases, allopathic) effect of the rapid initial growth and exuberant development of the biomass.
- The growth and decomposition of the green manure stimulates many species of micro-organisms, mainly soil micro-organisms, whose activity improves the physical and chemical dynamic soil properties.
- It offers multiple uses on the farm. Some types of green manure have a high nutritional quality and can be used for animal feed (oats, peas, pigeon peas and lablab), for human consumption (lupine and pigeon peas), or even as a source of firewood (*leucaena*).

CHARACTERISTICS TO CONSIDER WHEN SELECTING TYPES OF GREEN MANURE

The main characteristics that should be considered when selecting types of green manure are as follows (Amado and Wildner, 1991):

1. Rapid and aggressive initial growth and efficient soil cover.
2. Producing large amounts of biomass (green and dry matter).
3. Capacity to recycle nutrients.
4. Ease of establishment and management in the field.
5. Resistance to attacks by pests and diseases and not act as a host.
6. A deep penetrating and well developed root system.
7. Easy to manage when being incorporated into the soil and during establishment of the crops.
8. Potential for multipurpose use on the farm.
9. Tolerance or resistance to drought and/or frosts.
10. Tolerance to low soil fertility and be adapted to degraded soils.
11. Ability to produce seeds in sufficient quantities to increase the areas under the crop.

12. Not invading, so causing difficulties for the succeeding crop or the crop rotation.

Other characteristics, suggested by Malavota·(1967, cited by Muzilli *et al.*, 1980):

13. Members of the legume family.
14. Seeds of medium size (1 000 to 1 500 seeds/kg), which are able to germinate in soil prepared conventionally.
15. Robust species able to resist inclement weather.
16. Seeds that are permeable to water so that germination is facilitated.
17. Not climbing, particularly if a perennial.

Other characteristics can usefully include:

18. Readily adaptable to the cropping systems in the region.
19. Capacity to re-sprout in the event that parts of the area are cut down.
20. Naturally reseeding.

This long list does not imply that each species should have all the characteristics. In reality, some of these can be discarded, depending on the species of the green manure (winter or summer; bush or creeper; short or long cycle), the cropping system and the conditions of the farmer. For this reason and within the conditions of the particular farm, only some of the characteristics will be of fundamental importance and so the list serves more as a guide to selection criteria.

Main species used as green manure

The main species used as green manure and soil cover in Brazil and other Latin American countries are shown in Table 27.

CHARACTERISTICS OF GREEN MANURE

Spring/summer green manure

In the Southern Hemisphere, green manure is sown during the period of October to January. The most commonly used species are legumes such as velvet beans, jack beans, lupine and the crotalarias, amongst others. Their main advantages include the large production of biomass, the high amount of nitrogen that is fixed biologically and the soil cover during the season of high rainfall. The major disadvantage of the Spring-Summer green manures is occupying the soil during the season for the main economic Summer crops. In order to diminish this inconvenience, it is recommended that the farm be subdivided into parts where the green manure will be used in various stages.

Autumn/winter green manure

This method plans the use of green manure during the winter season, generally during the seasonal break between the main commercial crops. Because vast areas in the South of Brazil remain fallow during the winter season, subject to erosion, leaching of nutrients and proliferation of weeds, the method has been rapidly diffused amongst farmers.

TABLE 27
List of the main species used as green manure and soil cover

Scientific name	Common name		
	Spanish	English	Portuguese
1. *Avena byzantina* C. Koch	Avena amarilla	Algenina oat, red oat.	Aveia vermelha
2. *Avena sativa* L.	Avena blanca, avena común, avena	Oat, common oat.	Aveia
3. *Avena strigosa* Schreb	Avena negra, avena brasileña	Naked oat, sand oat.	
4. *Cajanus cajan* (L.) Milsp.	Gandul, guisante de Angola, frijol de árbol, frijol de Congo, frijol de palo.	Cajan pea, pigeon pea, gungo pea, Angola pea, red gram, (India: Dhall Toor).	Ervilha de Angola, guandú.
5. *Canavalia ensiformis* (L.) D.C.	Canavalia, frijol de puerco, frijol de espada, haba de caballo, guisante sable, poroto gigante.	Jack bean, Chickasaw bean, C. lima, horse bean, Patagonian bean, knife bean, overlook bean, sword bean.	Fava brava, feijâo sabre, feijâo de espada.
6. *Crotalaria grantiana* Harv.	Crotalaria	Crotalaria	Crotalária
7. *Crotalaria juncea* L.	Crotalaria, cáñamo de la India.	Bengal hemp, sunn-hemp.	Cânhamo da India, crotalária.
8. *Crotalaria mucronata* Desv.	Cascabel, guisante de cascabel, crotalaria, matraca.	Striped crotalaria, streaked crotalaria.	Casacavelheira, xique-xique.
9. *Crotalaria retusa* L.	Cascabel fétido, cascabelillo, maromera.	Devil bean, wedge-leaved crotalaria.	Chocalho, guizo de cascabel.
10. *Crotalaria spectabilis* Roth	Crotalaria, guisante de cascabel.	Show crotalaria.	Crotalária.
11. *Fagopyrum sagittatum Gilib.*	Alforfón, trigo sarraceno, trigo negro.	Beech wheat, buckwheat, Japanese buckwheat, silver hull.	Fagópiro, trigo sarraceno
12. *Lathyrus hirsutus* L.	Chícharo, guija velluda	Caley pea, hairy bitter vetch, singletary pea.	
13. *Lathyrus cicera* L.	Chícharo, lenteja forrajera	Chickling vetch, falcon pea, flat-pod pea vine.	Araca, chicharo branco
14. *Lathyrus sativus* L.	Chícharo, Chícharo común	Chickling pea, grass pea-vine, grass-pea.	Chícaro, chícharo comum.
15. *Lens culinaris Medik*	Lenteja	Lentil	Lentilha
16. Lolium multiflorum Lam	Raygrass, raygrass común, raygrass italiano, ballico	Annual ryegrass, Italian ryegrass.	Azévem, azévem italiano
17. *Lupinus albus* L	Altramuz, lupino, chocho	White lupine	Lupino branco.

Scientific name	Common name		
	Spanish	English	Portuguese
18. *Lupinus angustifolius* L.	Altramuz azul, lupino azul.	Blue lupine, narrow leafed lupine, New Zealand blue lupine.	Tremoço azul.
19. *Lupinus luteus* L.	Altramuz amarillo.	Yellow lupine, European yellow lupine.	Tremoceiro amarelo.
20. *Melilotus albus* Medik	Trébol dulce, trébol blanco dulce, trébol blanco de olor, meliloto.	Sweetclover, common sweetclover, honey clover, melilot.	Meliloto branco, trêvo doce.
21. *Melilotus officinalis* (L) Pall.	Meliloto, trébol de olor, trébol de olor, amarillo.	Common sweetclover, field melilot, yellow melilot.	Corva de rei, meliloto.
22. *Melilotus indicus* All.	Trébol de olor, alfalfilla, trevillo.	Indian clover, Indian melilot, yellow annual sweetclover.	Anafe menor, trêvo de cheiro.
23. *Ornithopus sativus* Brot.	Serradela.	serradela, bird's foot.	Serradela.
24. *Phacelia tanacefolia* Benth.	Facelia	Bluebell, California bluebell, scorpion weed.	
25. *Pisum sativum* (L.),	Arveja forrajera, arveja de campo, arveja verde de los campos.	Garden pea, field pea, grey field pea, Australian winter pea.	Ervilha, ervilha brava.
26. *Raphanus sativus* L. *var, oleiferus Pers.*	Rábano forrajero, rábano oleoginoso	Radish fodder, Japanese radish.	Nabo chinês, rábano.
27. *Secale cereale* L.	Centeno	Rye, common rye.	Centeio.
28. *Sesbania cannabina (Retz.)* Pers.	Sesbania	Sesbania.	
29. *Sesbania exaltata (Raf.) Cory.*	Sesbania común.	Hemp sesbania, sesbania.	
30. *Sesbania speciosa* Taub.	Sesbania	Sesbania.	
31. *Spergula arvensis* L.	Espérgola, pegapinto, esparcilla.	Corn spurry, corn spurry.	Cassamelo, espergula, gorga.
32. *Mucuna aterrima* (Piper et Trary) Merr.	Mucuna negra, frijol velludo.	Bengal bean, black velvet bean, Mauritius bean.	
33. *Mucuna pruriens* (L.) DC.	Mucuna rayada, frijol aterciopelado, frijol aterciopelado de Florida.	Florida velvet bean, deering velvet bean.	Café de Mato Grosso, para café.
34. *Trifolium incarnatum* L.	Trébol francés, trébol encarnado	French clover, crimson clover, carnation clover.	Erva do amor, trêvo encarnado.
35. *Trifolium subterraneum* L.	Trébol subterráneo.	Subterraneum clover, subclover	Trêvo subterrâneo.
36. *Trigonella foenum-graecum L.*	Heno griego, fenogreco	Fenugreek, fenugrec.	Ferro grego.

Scientific name	Common name		
	Spanish	English	Portuguese
37. *Vicia angustifolia* L.	Alverilla, veza de hoja angosta	Angusta vetch, narrow-leaf vetch.	
38. *Vicia articulata* Horn.	Arvejilla parda, garrubia	Bard vetch, monantha vetch, one-flowered tare.	Algarroba, ervilhaca parda.
39. *Vicia bengalensis L.*	Veza arvejilla.	Purple vetch.	
40. *Vicia ervilia* L. Willd.	Arvejilla amarga, yeros.	Bitter vetch, ervil, lentil vetch.	Ervilha de pombo, gêro.
41. *Vicia faba* L.	Haba común, haba, frijol de caballo, frijol forrajero.	Broad bean, great field bean, faba bean, pigeon bean, horse bean, marsh bean, small field bean.	Fava, fava do campo.
42. *Vicia hirsuta* L. S.F. Gray.	Arvejilla hirsuta.	Tiny vetch, hairy vetch.	Ligerâo, unhas de gato.
43. *Vicia pannonica* Crantz.	Arvejilla húngara.	Hungarian vetch.	
44. *Vicia sativa L.*	Arvejilla común, veza común.	Common vetch, vetch, golden tare.	Ervilhaca
45. *Vicia villosa* Roth	Arvejilla peluda, arvejilla de las arenas.	Hairy vetch, winter vetch, Russian vetch.	Ervilhaca das areias, ervilhaca peluda.
46. *Vigna sinensis* L. Walp.	Caupí, frijol de vaca, frijol chino, frijol de cuerno.	Cowpea, Cuba pea, long bean, common cowpea, blackeye cowpea.	Cultita, feijâo de liba.

Green manure mixed in association with the crops

In this method, the green manure is sown between the rows of the commercial crop and is particularly well adapted to situations where the soil has to be used as intensively as possible. Amongst the types of green manure most commonly used for this method are velvet beans associated with maize, perennial soybean grown amongst citrus fruits and vetch sown amongst grapevines. This type of green manure should be treated with caution in order to avoid competition with the commercial crop and consequently lower yields. The main advantages of the system are its intensive use of the soil, the efficient erosion control and the reduction in weed propagation.

Perennial green manure in areas under fallow

Green manure can be recommended in areas degraded due to management practices or in areas temporarily lying fallow.

The main species used include lupine and species such as *Indigofera, Leucaena, Tephrosia*, and *Crotalaria,* amongst others. These species, due to their deep root system and high biomass production, have a double advantage in recovering the soil properties and allowing their use as animal feed.

MANAGEMENT OF GREEN MANURE

Factors to be considered for establishing green manure crops

In order for the green manure to express its biomass production potential to a maximum, it is necessary that minimum conditions favouring its growth and development be offered. It is fundamental to understand the crop needs as these concern temperature, soils and moisture availability (Bulisani and Roston, 1993). These three parameters allow recognition of the behaviour of the particular green manure, definition of the optimum sowing dates and the best regions for establishing the crop, all conditioned by the particular soil type.

As far as temperature requirements are concerned, green manure can be divided into two main groups: green manure for the subtropical and temperate regions and green manure for tropical regions. These are most commonly referred to as green manure for winter or summer.

The winter species are suitable for the period during the year when the high summer temperatures start dropping and in particular, lower night-time temperatures start to appear. Sowing should be undertaken in such a way as not to prejudice the vegetative growth or the reproductive phase that occurs at the start of Spring.

In the case of tropical or summer species, it is necessary to be wary of low temperatures occurring at the start of the growing season. These could cause irreversible damage to the green manure by retarding the growth, perhaps making reproduction impossible (flower abortions, fruit burn-up), or restricting biomass production. The date chosen for sowing summer-season green manure determines the final crop height and the production level of the biomass from upstanding species, or the lateral expansion of creepers. In this way, once the rainy season has commenced, later sowing of lupines and crotalarias causes several adverse effects (Wutke, 1993), (Wildner and Massignam, 1994a, b, c). Crop height may be reduced; in one case this decreased from 3 m to 1.0-1.5 m with a corresponding reduction in biomass production (Wutke, 1993; Wildner and Massignam, 1993). Soil cover may be reduced, the incidence of pests may increase and grain harvest may either be eased or made more difficult.

The most commonly used types of green manure have a wide adaptability to different soil types. In general, the legumes require a minimum of soil fertility, which translates into an adequate availability of Ca, Mg, P and K (Bulisani and Roston, 1993). Some legumes are more tolerant to degraded soil conditions, amongst them lupine and various crotalarias. Summer legumes, on the other hand, seem to be less needy of soil fertility than winter legumes. Other graminaceous, cruciferous and cariophilaceous species are even less exigent than the legumes.

Soil moisture availability, represented by its amount and distribution, markedly influences the development of green manure and the determination of its optimum sowing date. It is important, therefore, to identify the periods of critical moisture deficit so that sowing dates may either be brought forward or delayed. The most critical phase for establishing legumes is determined by the date of germination and emergence of the plants when a lack of moisture can restrict the development of an adequate plant population (Bulisani and Roston, 1993). During the following phases of the vegetative cycle, due to the nature of the root system, the moisture requirements and the particular stage of plant development, the effects of moisture deficiency are less pronounced.

Management of the biomass

The amount of biomass produced in a particular area of the farm depends basically on the interest and the aims of the farmer. The time during which the vegetative cover will remain is determined by considering the particular production system adopted on the farm and can be either longer or shorter than that normally recommended for this particular conservation practice. The soil cover when being cropped, should not be removed during any period of the year if the physical, chemical and biological integrity of the soil is to be maintained (Wutke, 1993).

The farmer may opt for one of three basic management systems:

1. Total incorporation of the biomass, characterized by the traditional type of green manure.
2. Partial incorporation of the biomass, characterized by the so called minimum tillage.
3. Management of the biomass, without incorporating it into the soil, characterized by direct drilling.

Total incorporation of the biomass: is the practice that is best known and accepted by farmers. The incorporation can be realized at any moment, depending upon the farmer's objectives. The season traditionally recommended for this is during full flowering of the green manure. This is the phase that corresponds to the maximum accumulation of biomass and nutrients. If incorporation is done earlier, the rate of decomposition of the biomass will be greater but the nutrient levels will be less. When incorporation is delayed, the plants become more fibrous (a higher C/N ratio) and the rate of decomposition is slower. The decisive factor for this operation or for the timing of the moment for establishing the green manure, will depend mainly on the sowing season for the following crop. The tillage operations should be carried out with ploughs and discs (for levelling or ridging).

Partial incorporation of the biomass: a minimum of soil preparation operations should be used for the partial incorporation of the biomass so as to provide favourable conditions for seed germination and establishment of the crop (Curi *et al.* 1993). For the conditions of the small farmers in the South of Brazil, the idea of minimum tillage with *vicia sativa* as a winter cover crop has been introduced, using animal traction (Monegat, 1981). In this case, the only operation for soil preparation consists of opening up a furrow at the spacing to be used by the following crop, so that the soil between the rows remains protected. In this system, the vegetative cover is partially incorporated during the ploughing, generally between 20 and 40 percent. The rest of the cover can be maintained on the surface or be totally or partially incorporated during the application of nitrogenous fertilizer or during the weed control operations (Monegat, 1991). The system is viable in areas where there are few weeds. If this is not the case, chemical weed control may be practised. In addition to various other benefits that it offers, the minimum tillage also reduces the labour requirements for establishing the crop, compared to conventional tillage.

Minimum tillage with animal traction can be undertaken in four different ways, depending on the species and the phase of the vegetative cycle that has been reached by the green manure (Monegat, 1991):

a. *Minimum tillage before flowering of the green manure:* this is carried out in areas cultivated with low-standing or creeper type green manure which has a slow initial development rate (serradela or bird's foot, purple clover, lentil), low biomass production and a long growth cycle. The management is undertaken when there is complete soil cover. The furrows should be wide and sowing should preferably be along parallel rows. This system allows early sowing of maize and natural reseeding of the green manure. At the end of the cycle of the cover crop and between the rows of the main crop, it is possible to carry out a second and later direct sowing which thus provides a substitution type of association or a succession of crops.

b. *Minimum tillage during the phase of full flowering of the green manure:* in this case the management is undertaken during the stage of full flowering of the green manure. On many occasions, when there is a large production of biomass, there can be difficulties to open up the furrows. Sometimes the plough becomes blocked and on other occasions, the green manure falls into the open furrow, making sowing and the eventual plant emergence difficult. In order to avoid these problems, it is recommended to open up the furrows as soon as there is 100 percent soil cover, an operation also known as pre-furrowing. A plough with a medium or large-sized body should be used. Pre-furrowing delays the growth of the green manure and avoids the production of an excessive amount of biomass. During the flowering stage, the second and final furrowing operation is carried out. This type of minimum tillage is used for common vetch, hairy vetch, chickling vetch and other similar species. At the end of the vegetative cycle, it is also possible to establish another crop by direct drilling, either in association or as a following crop and between the rows of the main crop.

c. *Minimum tillage after the harvest of winter cereals:* in this system, immediately after harvesting the winter cereals (wheat, triticale, oats, rye), the area is furrowed and the main crop sown. Weed control is done in a similar way to that in the conventional system.

d. *Minimum tillage after flattening the green manure:* this is the typical example of minimum tillage for summer green manure, such as velvet beans, but it can also be used with winter green manure. The plants are flattened with a knife-roller, a disc roller or even a disc plough or a hand scythe or a mechanical mower. Chemical desiccation can also be used. One or two weeks after flattening the green manure and when the biomass is well dried, the furrowing operation is undertaken. In the case of straight-standing green manure crops such as oats, rye, forage turnips, crotalarias, etc., the use of a traditional ridging plough is recommended. When green manure that forms vines is present such as common vetch, hairy vetch or velvet beans, it is recommended to adapt an indented cutting disc on the front of the plough to cut the plant stems.

It is important to point out that minimum tillage with animal traction is a system adapted for crops which are sown at a wide row spacing (close to one metre between rows) such as maize, cassava and tobacco.

It should also be indicated that minimum tillage has some disadvantages such as:

- Greater difficulty to make the furrows than in conventional tillage.
- Greater presence of rats and soil diseases.
- It is not recommended for areas heavily infested with weeds.

- If it is not well planned, the system can interfere adversely with the traditional production systems (inter-cropping and substitution associations) (Monegat, 1991).

Motorized minimum tillage is not a system that is widely known and it is only used in the South of Brazil for onion production. A small strip of soil is prepared for each row of the crop (onions) and fertilizer is applied at the same time. The soil preparation is done with specially adapted "Rotacaster" seed drills, pulled by medium-powered tractors (Silva *et al.*, 1993a).

Management of the biomass without incorporation into the soil. The sequence of operations starts with management of the biomass without incorporating it into the soils, and ends by direct drilling the main crop into the un-tilled land. Special direct drills are used which open up a small slot of a depth and width which is just sufficient to guarantee good covering and soil contact with the seed (about 25 to 30 percent of the soil surface is tilled) (Curi *et al.*, 1993). Normally, weed control for this system is done chemically.

Management of the biomass can be done with mechanical methods (knife rollers or in special cases, mowers or choppers) or by chemical desiccation with herbicides. The mechanical methods should be used with considerable care, mainly as this concerns timeliness, so as to avoid problems such are poor flattening or re-sprouting. For this reason, crop flattening should be done during full flowering of the green manure or when the grains are milky, depending on the species being managed. As regards the chemical method, total action products (desiccants) are normally used, taking care not to contaminate and damage the environment. The range of types of direct drills is very great. There are specially adapted hand-operated planters (matracas, pole-sticks or the "saraquá" punch-planter). There are also tractor-operated machines with precision drills and with electronic metering mechanisms. The technology of direct drilling for sowing large areas offers many alternatives, whilst for small farmers, there is still a need for more and better equipment. Some technical and functional problems have been reported concerning the use of the "saraquá" punch planter for direct seeding of maize, beans and soybean (Monegat, 1991):

- Difficulty in lining up the seed rows.
- Difficulty in penetration by the "saraquá" when the layer of green manure is thick or the soil is dry or compacted.
- Uneven seed germination and plant development during periods of moisture stress, due to the seeds being placed near the surface.
- Chlorosis of plants indicating nutritional problems (nitrogen deficiency).
- Etiolation when the humus layer is thick.
- Frost effects can be accentuated due to the dead covering matter insulating the young plants from the warm soil.

EFFECTS OF GREEN MANURE ON THE SOIL PROPERTIES

Effects on the soil physical properties

The effects of green manure or soil cover can be observed during two stages (Muzilli *et al*,. 1980):

a. The first stage refers to the protection of the surface layers of the soil by the plants.
b. The second stage refers to the incorporation of the vegetative matter into the soil.

The vegetative cover, whether live or dead material, is the single most important factor influencing the topsoil due to preventing the break-up of the aggregates and the formation of surface crusts, which would reduce water infiltration (Amado, 1985). It also reduces runoff, the concentration and the size of the transported sediment particles and thus, the rates of loss of both soil and moisture.

The vegetative cover also influences the soil moisture and temperature. The reduction of moisture losses can be attributed to the combination of several factors. There are noticeable reductions in the rates of surface evaporation and of surface runoff and increases in infiltration rates and moisture retention capacity of the soil (Moody, 1961 and Eltz *et al.*, 1984, cited by Amado *et al.*, 1990). The difference in the soil moisture content is more pronounced during periods of drought, providing evidence that the soil cover shortens the periods of moisture deficits (Amado *et al.*, 1990).

FIGURE 32
Variation of the maximum and minimum temperatures over 5-day periods at a 5 cm depth in the soil during a trial of vegetative matter incorporation and soil cover before sowing beans in Campinas, SP. (Miyasaka, 1984)

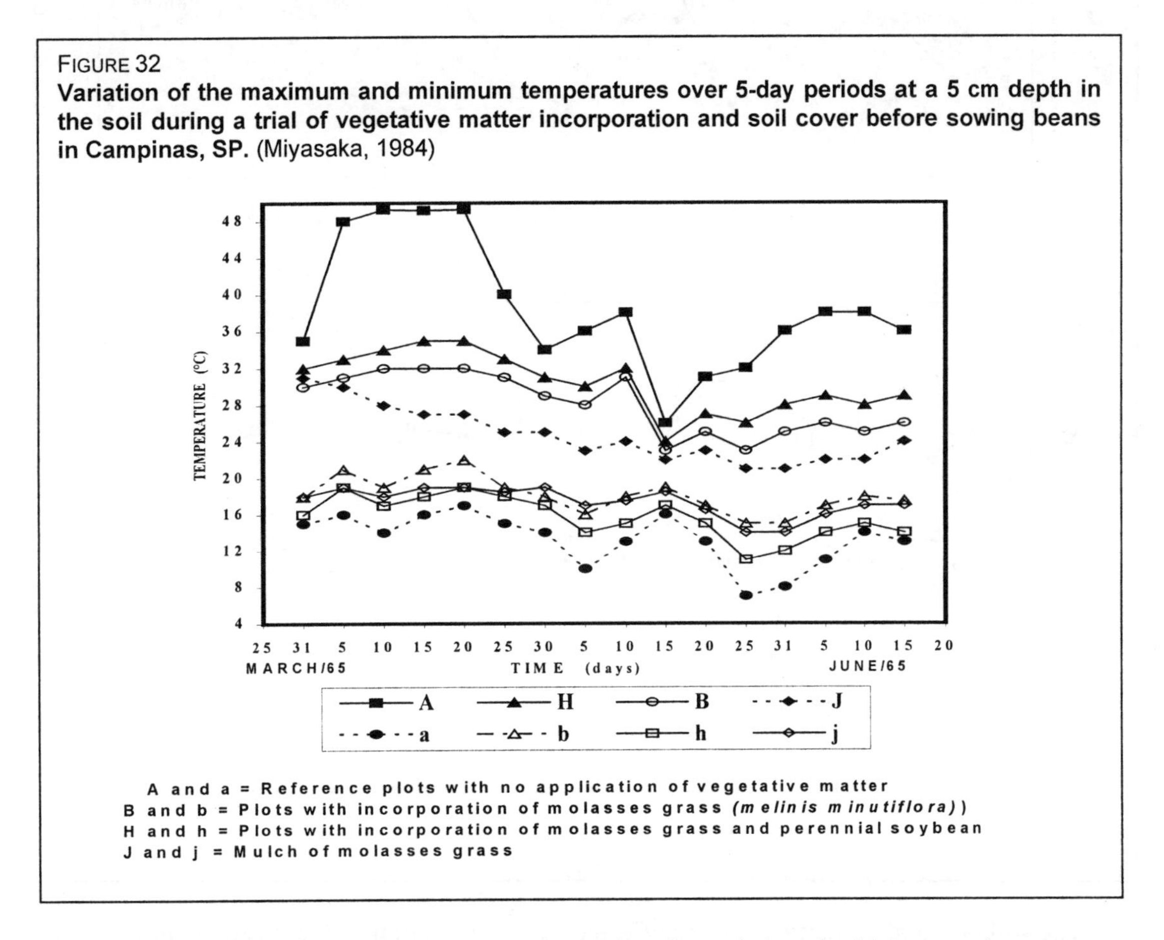

Studies of soil management systems for maize showed that the highest moisture content was registered in plots with a cover of naked oats and the lowest, when covered with chickling vetch (Derpsch *et al.*, 1985). The soil moisture content in plots covered with naked oats was from 3 to 7.4 percent higher than in the uncovered plot. In general, the results showed clearly that the losses of soil water were reduced by the presence of plant residues grown over the winter and maintained on the soil surface (Figures 32 and 33). It is important to note that the analysis of the maximum and minimum soil temperatures is of fundamental importance because their effects on the biological activity, seed germination, root growth and absorption of ions.

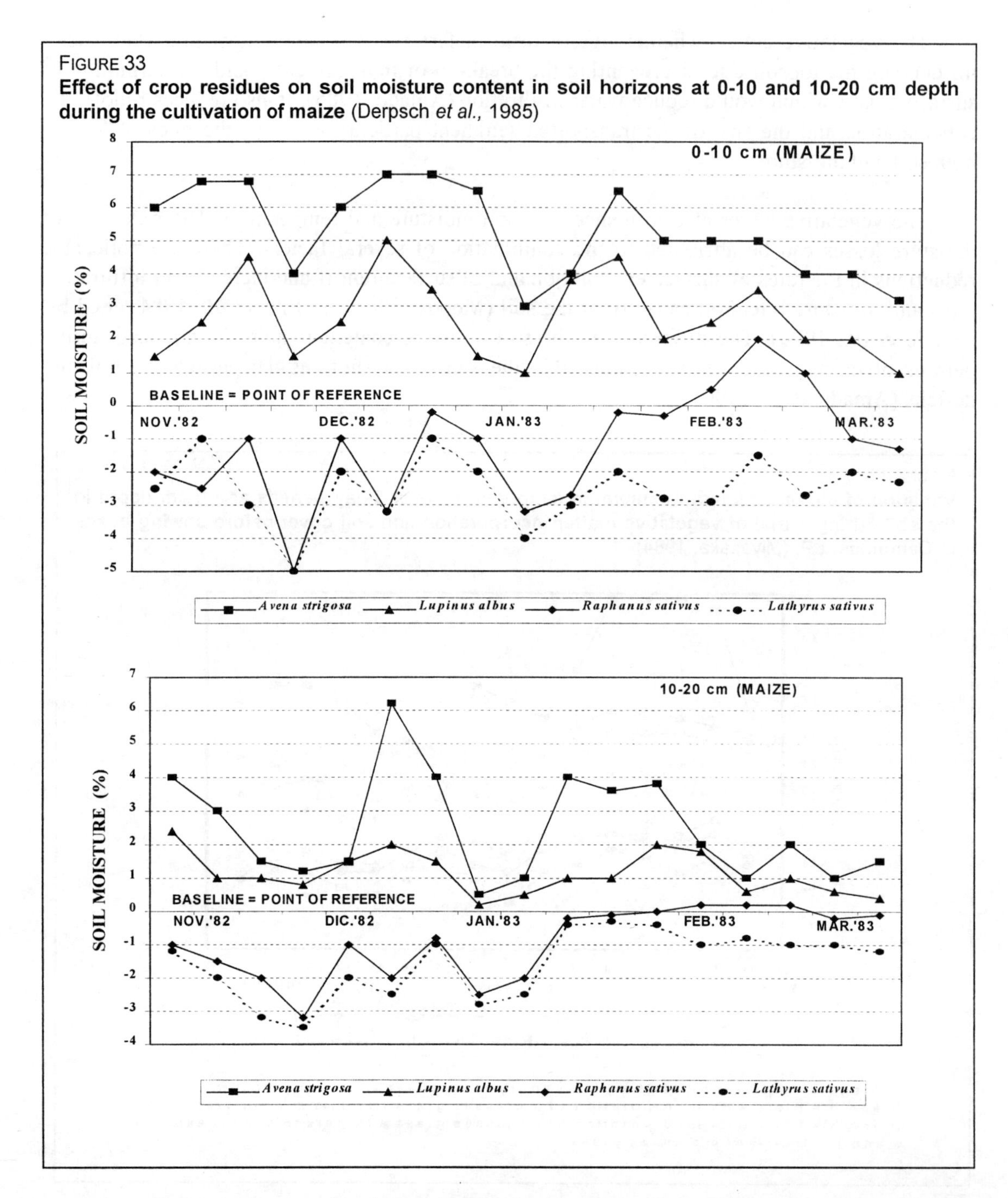

FIGURE 33
Effect of crop residues on soil moisture content in soil horizons at 0-10 and 10-20 cm depth during the cultivation of maize (Derpsch *et al.*, 1985)

The soil physical properties that are affected by incorporation of the green manure include the structure, moisture retention capacity, consistency and density. Other properties such as the porosity, aeration, conductivity, hydraulics and infiltration are allied to the modifications to the soil structure. However, this effect depends on climatic factors and the soil characteristics (Muzilli *et al.*, 1980).

The water infiltration rates in a soil, after having a green cover, were evaluated with the method of concentric rings (Derpsch, 1984). The results showed that they increased by up to 416 percent in a Dystrophic Red Latosol (FAO: Rhodic Ferralsol) and up to 628 percent is as structured Red Earth (FAO: Nitosol), when compared with a plot of wheat. Furthermore, the increased infiltration persisted until the next crop of soybean was established.

Studies were made of the infiltration rate as influenced by different types of soil cover, with a rainfall simulator (Sidiras and Roth, 1984). The highest infiltration rate was observed when there was the largest amount and best quality of green manure. The results in this study revealed less significant differences than those recorded in the work of Derpsch (1984) due mainly to the different methodology adopted.

Derpsch (1984), also showed that when comparing conventional tillage of plots for wheat with plots covered with green manure, the cover had a positive effect on the soil consistency. According to the author, the green cover leaves the soil very friable due to the biological tillage of the soil.

Effects on the soil chemical properties

Depending on the quality and quantity of green matter production, various species of green manure can bring about the recycling of nutrients, supply nitrogen and maintain or increase the level of organic matter content in the soil. According to Muzilli *et al.*(1980), the contribution of the green manure to improvement of the organic matter content depends on the amount of residues incorporated, the frequency of the incorporation and the quality of the material.

The nutrient recycling capacity is recognized through many studies. The recycling of the nutrients can occur under four different situations:

1. Deep-rooting green manures being nutrients leached towards the deeper soil layers, beyond the reach of the roots of the commercial crops, towards the surface. After the incorporation of the green manure and mineralization of its organic matter, the nutrients again become available to the crop.
2. The nutrients located in the plough layer of the soil are incorporated into the vegetative weaving of the green manure and are protected from possibly being removed by erosion. After incorporation and mineralization of the green manure's biomass, the nutrients becomes available again to the plants.
3. The legumes, through a process of symbiosis with the *Rhizobium*, encourage the fixing of atmospheric nitrogen. It is possible that non-leguminous species have similar or even higher concentrations of nitrogen than those in the legumes. In this case, there is excellent advantage taken of the natural nitrogen in the soil once it is determined that there is no characteristic symbiosis.
4. The transformation of the nutrients found in a form that is not available to the plants into one that is easily assimilated (an association with mycorrhiza).

Results obtained by Wildner (1990) showed the recycling capacity of different species used as winter and summer green manure (Tables 28, 29 and 30). Work undertaken at the Ituporanga

Experiment Station in the region of the high valley of Itajaí, confirmed these previously mentioned results (Amado, 1991).

TABLE 28

Production of biomass and analysis of the nutrients in the vegetative cover of winter green manure species evaluated at the CPPP. Chapecó, SC, 1990 [1]

Species	Matter (t/ha)		Nutrients (%)					Organic carbon (%)	C/N ratio[3]
	Green	Dry[2]	N	P	K	Ca	Mg		
Naked oats (*Avena strigosa*)	31.5	7.7	1.39	0.17	2.30	0.38	0.17	37.6	27.1
Rye (*Secale cereale*)	35.4	6.2	0.97	0.20	2.05	0.32	0.10	39.3	40.5
Ryegrass (*Lolium multiflorum*)	29.8	4.8	1.01	0.13	2.61	0.52	0.18	36.4	36.1
Chickling pea (*Latyrus sativus*)	30.8	3.9	2.70	0.26	2.74	0.56	0.28	38.7	14.3
Vetch (*Vicia sativa*)	18.9	3.6	3.00	0.31	2.51	1.08	0.30	36.5	12.1
Hairy vetch (*Vicia vilosa*)	23.9	5.0	3.41	0.35	2.98	0.90	0.26	37.9	11.1
Forage oats (*Pisum arvense*)	28.9	3.0	2.89	0.32	2.44	0.82	0.28	38.8	13.4
Forage turnip (*Raphanus sativus*)	28.0	3.5	2.32	0.35	3.59	2.09	0.38	32.7	14.1
Corn spurry (*Spergula arvensis*)	33.0	3.8	1.62	0.30	2.90	0.44	0.74	36.9	22.8

Observations:

[1] The data presented refer to evaluations made in 1985, 1986 and 1987.

[2] Dry matter after oven drying at 60°C.

[3] Ratio of organic C to total N.

Source: Wildner, 1990.

TABLE 29

Nutrient content of the components (stems and leaves) of summer cycle annual species with a potential for use as a green manure, soil cover and for soil recovery. Chapecó, SC, 1990[1]

Species	Stems						Leaves					
	MS[2] (t/ha)	N (%)	P (%)	K (%)	Ca (%)	Mg (%)	MS[2] (t/ha)	N (%)	P (%)	K (%)	Ca (%)	Mg (%)
Crotalaria juncea (Crotalaria/USA)	10.8	1.12	0.07	1.01	0.41	0.24	2.5	2.73	0.24	1.40	1.89	0.72
Crotalaria retusa (Crotalaria/IAC)	3.0	1.52	0.14	2.25	0.61	0.45	2.0	3.39	0.22	1.47	2.54	0.66
Crotalaria spectabilis (Crotalaria/IAC)	7.5	1.52	0.11	2.49	0.70	0.28	2.5	3.72	0.24	1.96	1.97	0.44
Crotalaria lanceolata (Crotalaria/SLO)	4.1	1.53	0.09	2.22	0.32	0.30	2.3	5.00	0.23	2.02	1.01	0.43
Crotalaria grantiana (Crotalaria/IAC)	5.6	0.97	0.05	1.42	0.50	0.11	2.1	4.60	0.26	1.80	1.33	0.36
Canavalia ensiformes (frijol de puerco/USA)	3.9	1.40	0.18	1.89	0.60	0.25	3.2	3.82	0.22	1.91	2.56	0.55
Cajanus cajan (gandul/CNPAF)	6.0	1.39	0.10	1.01	0.47	0.18	1.7	3.86	0.23	1.73	1.07	0.38
Stizolobium deeringianum (Mucuna/USA)	2.2	1.97	0.16	2.37	1.33	0.32	1.3	4.00	0.24	1.28	1.81	0.37
Stizolobium sp. (Mucuna rayada/USA)	4.5	1.78	0.12	2.03	0.98	0.36	3.2	4.15	0.26	1.25	2.04	0.39
Stizolobium niveum (Mucuna ceniza/USA)	4.3	1.72	0.19	1.46	0.77	0.34	2.2	4.41	0.33	1.16	1.50	0.47
Stizolobium aterrium (Mucuna negra/IAC)	3.6	2.24	0.19	1.96	0.77	0.22	2.4	4.39	0.29	1.10	1.58	0.41

Observations

[1] The data presented refer to evaluations made during the harvests of 1986/87, 1987/88 and 1988/89.

[2] Dry matter after oven drying at 60°C.

Source: Wildner, 1990.

TABLE 30
Nutrient content of the stems and leaves of winter cycle, semi-perennial and perennial species with a potential for use as a green manure, soil cover and for soil recovery. Chapecó, SC, 1991[1]

Species	Stems						Leaves					
	MS[2] (t/ha)	N (%)	P (%)	K (%)	Ca (%)	Mg (%)	MS[2] (t/ha)	N (%)	P (%)	K (%)	Ca (%)	Mg (%)
Crotalaria paulina	13.2	1.29	0.08	1.85	0.98	0.39	3.7	3.56	0.25	2.04	2.41	0.58
Crotalaria mucronata	10.0	1.40	0.08	1.99	0.51	0.25	3.1	5.80	0.24	1.31	1.46	0.47
Anileira	7.9	1.32	0.11	1.01	0.82	0.23	1.9	4.77	0.29	1.72	1.18	0.47
Leucaena/EMPACA	3.8	1.35	0.08	0.92	0.53	0.26	2.2	3.78	0.17	1.12	1.13	0.70
Leucaena/CV PERU	10.5	1.54	0.09	1.29	0.39	0.26	5.1	3.60	0.16	1.49	0.95	0.45
Pigeon pea/CV KAKI	17.0	1.54	0.10	1.08	0.59	0.26	4.0	4.08	0.26	1.36	0.87	0.32
Pigeon pea/CPPP	11.3	1.40	0.11	1.13	0.57	0.22	2.8	4.08	0.27	1.55	1.07	0.30
Tephrosia	6.0	1.95	0.12	1.08	1.08	0.26	1.7	4.06	0.22	1.22	1.01	0.25

Observations:
1 The data shown refer to evaluations undertaken during the harvests of 1986/87, 1987/88 and 1988/89.
2 Dry matter after oven drying at 60°C.
Source: Wildner, 1990.

Effects on the biological soil properties

Plants used as green manure, even before being managed in this manner, have an influence on the biological properties of the soil due to the physical effect of attenuating temperature variations and through maintaining good moisture conditions in the soil (Derpsch, 1984).

After management of the biomass, the presence of organic matter is the next most important factor influencing the activity and the population of the microorganisms because it is the organic matter that supplies the energy needed by the soil organisms. For this reason, the higher the production level of the biomass by the green manure, so will be the macro and microbial population of the soil (Figure 34).

FIGURE 34
Number of arthropods per 300 cm^3 soil samples on direct and conventional soya sowing after wheat and green cover cropping (adapted from Kemper and Derpsch, 1981). Source: Almeida, Siembra Directa IV(16): 4-5, 1985b

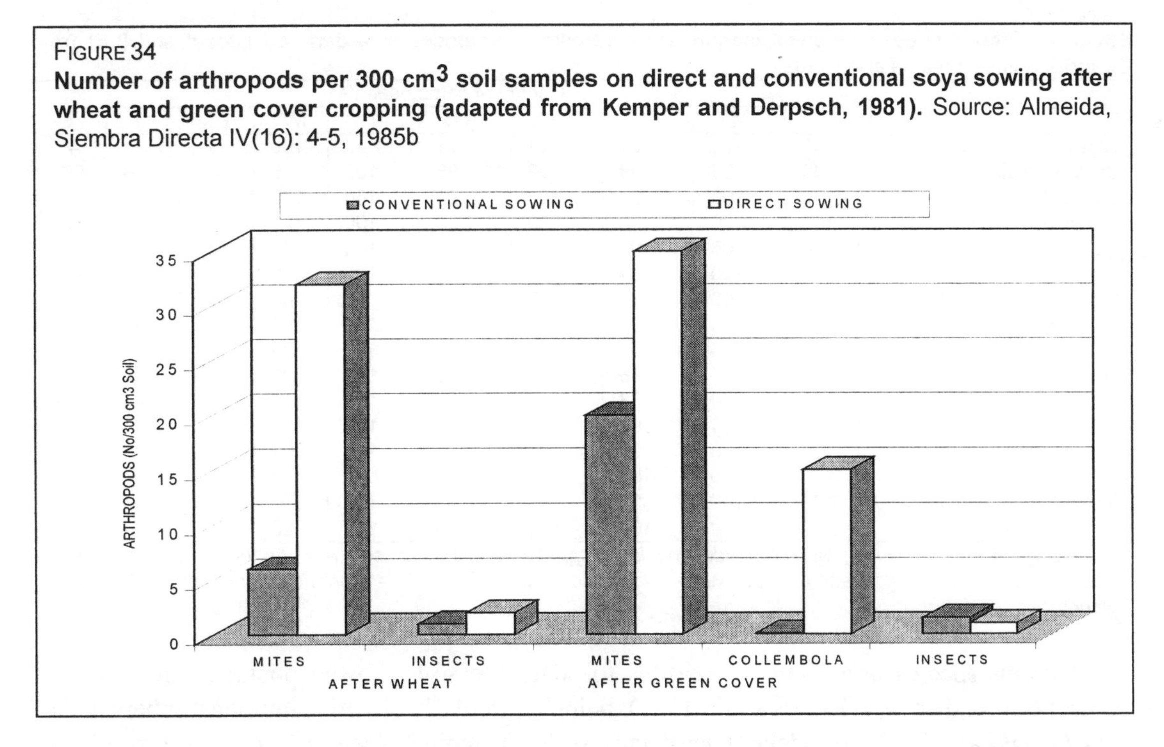

As the vegetative cover reduces, so soil movement increases and a reduction in the population of soil organisms is inevitable (Figure 35).

FIGURE 35
Influence of the intensity of soil disturbance on the population of soil organisms (Number of earthworms/m2) in the horizon of 0-10 cm (Adapted from Kemper and Derpsch, 1981). Source: Almeida, Siembra Directa IV(16): 4-5, 1985b

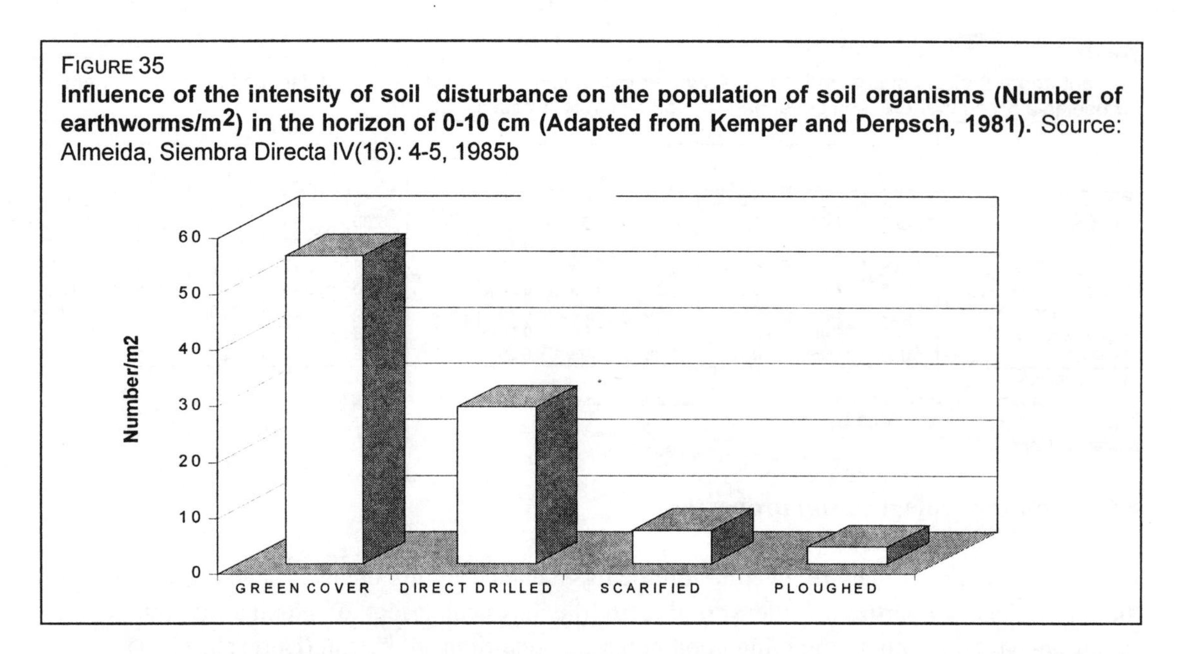

According to Almeida (1985b), soil preparation operations cause the death of the major part of the soil organisms, imposing conditions of high temperatures and alternating situations of drought and excessive moisture, which affect the soil organisms to a greater or lesser degree.

Sharma *et al.* (1982) and Santos *et al.* (1987) consider green manure as one of the most valuable and low-cost methods for the control of nematodes (Table 31).

TABLE 31
Effect of different species of green manure for controlling nematodes in a dark-red latosol soil (LE) from Cerrado. Source: Sharma *et al.*, 1982.

Species	Control of nematodes (%)									
	P	M	D	A	AA	T	MA	PT	O	S
Tagetes erecta	100	100	91	12	1	---	100	---	---	4
Cajanus cajan	100	96	98	92	98	100	75	---	---	96
Canavalia ensiformis	100	100	96	100	99	100	100	---	---	99
Crotalaria grantiana	100	100	100	100	100	100	100	100	100	99
Crotalaria juncea	100	100	100	81	97	100	30	---	---	96
Crotalaria paulina	100	100	94	94	99	100	100	---	---	97
Crotalaria spectabilis	100	100	94	93	100	100	100	---	100	97
Cyamopsis psoralioides	100	100	80	100	98	100	100	---	100	98
Dolichos lablab	91	99	94	100	100	100	100	---	100	98
Indigofera tinctoria	100	100	98	100	99	100	100	---	---	99
Phaseolus aurens	85	90	73	15	2	3	---	---	---	5
Sesbania aculeata	100	100	98	100	100	100	30	100	100	100
Stizolobium deeringianum	100	100	93	97	99	100	---	---	---	95
Stizolobium niveum	100	100	100	100	98	25	100	---	---	93
Stilozobium aterrium	100	100	93	93	99	100	100	---	100	95
Tephrosia candida	100	100	100	100	96	100	25	---	---	94

P = *Pratylenchus brachyurus;* M = *Meloydogyne javanica*; D = *Ditylenchus* sp.; A = *Aphelenchoides* sp.; AA = *Aphelenchus avena*; T = *Tylenchus* sp.; MA = *Macrosposthora ornata*; PT = *Paratrichodorus minor*, O = Other Tylenchytes; S = Saprophytes.

Various species used between seasons to cover the soil, showed positive effects on the control of root diseases (Santos *et al.*, 1987), being particularly effective, amongst others, naked oats (*Avena strigosa*), serradela (*Ornithopus sativus*), lupine (*Lupinus sp.*), ryegrass (*Linum usitatissimum*) and cabbage (*Brassica campestris*). Naked oats particularly stands out as an option for rotating crops in wheat-growing areas with phytopathelogic problems, for instance “take-all” (*Ophiobulus sp.*).

Effects of green manure on crop yields

Research studies have highlighted the marked influence on crop yields of green manure and residues of the plants used for soil cover. Similar results have been obtained for both winter and summer types of green manure.

Results obtained by Derpsch *et al.* (1985) demonstrated the marked influence of winter green manure on the production of maize, beans and soybean. The highest maize yields were obtained after white lupine (*Lupinus albus*) and vetch (*Vicia sativa*). For bean production, the best yields were achieved following forage turnips and naked oats. According to these authors, the number of grains per plant was the factor that most influenced the bean production level. Muzilli *et al.* (1983), recommended winter green manure such as white lupine as an alternative for recovering the productive capacity of soils degraded through intense usage. It also reduces the cost of nitrogenous fertilizers needed for maize cultivation. However, the needs of cultivars or hybrids can differ:

a. the response may be positive and the green manure is able to supply adequate nitrogen to the crop;

b. the response is positive but the practice of using green manure might not be sufficient to satisfy the nitrogen needs, particularly in cases of highly demanding cultivars and hybrids (for example, the AG-162 double hybrid).

Results obtained by Scherer and Baldissera (1988) in a characteristic soil from the basalt coasts of the western region of Santa Catarina, Brazil, showed the positive effects of velvet beans (*Stizolobium niveum*) as a green manure when intercropped with maize. In this case, greater benefits were achieved with the velvet beans when the plot was tilled conventionally rather than by direct drilling and with minimum tillage (Figure 36). The authors hypothesized that this effect was due to the greater amount of nitrogen originating from the mineralization of the organic compost and the increased freeing of this element over the short term. They also noticed that the increase in productivity encouraged by the use of velvet beans was equivalent to an application of 30 kg N/ha (700 kg/ha of maize), but that this difference diminished as the rate of nitrogenous application was increased.

Effects of green manure on weed control

Weed control is more efficient in cropping systems with a dead surface cover, particularly of winter species. The action of the dead cover is mainly due to the allopathic effect of the decomposition products (Lorenzi, 1984). Table 32 shows some of the most common examples of allopathy and plant incompatibility.

According to Almeida *et al.* (1984), the allopathic effects are specific and thus the complex that develops in the different dead covers is distinct both quantitatively and qualitatively and depends on the type of plant residue in the humus. According

TABLE 32

Allopathic effect of crops and species used for green manure (Bidens alba), on the germination of seeds from a selection of weeds.

Species	Treatment	Weeds
Naked oats	Dead cover	Plantaginea
Ryegrass	Fallow	Queensland hemp
Black velvet beans	Soil cover	Purple nutsedge and hairy beggar-ticks
Jack beans	Soil cover	Purple nutsedge
Sugar cane	Straw	Hairy beggar-ticks (*)
Crimson clover	Crimson clover	Auto-toxicity
Flax	Flax	Auto-toxicity
Rape (colza)	Rape (colza)	Auto-toxicity
Sunflower	Sunflower	Wild groundnut (**)

(*) The action of cane straw is so strong that it can affect the development of it re-sprouting.
(**) Practically insensitive.

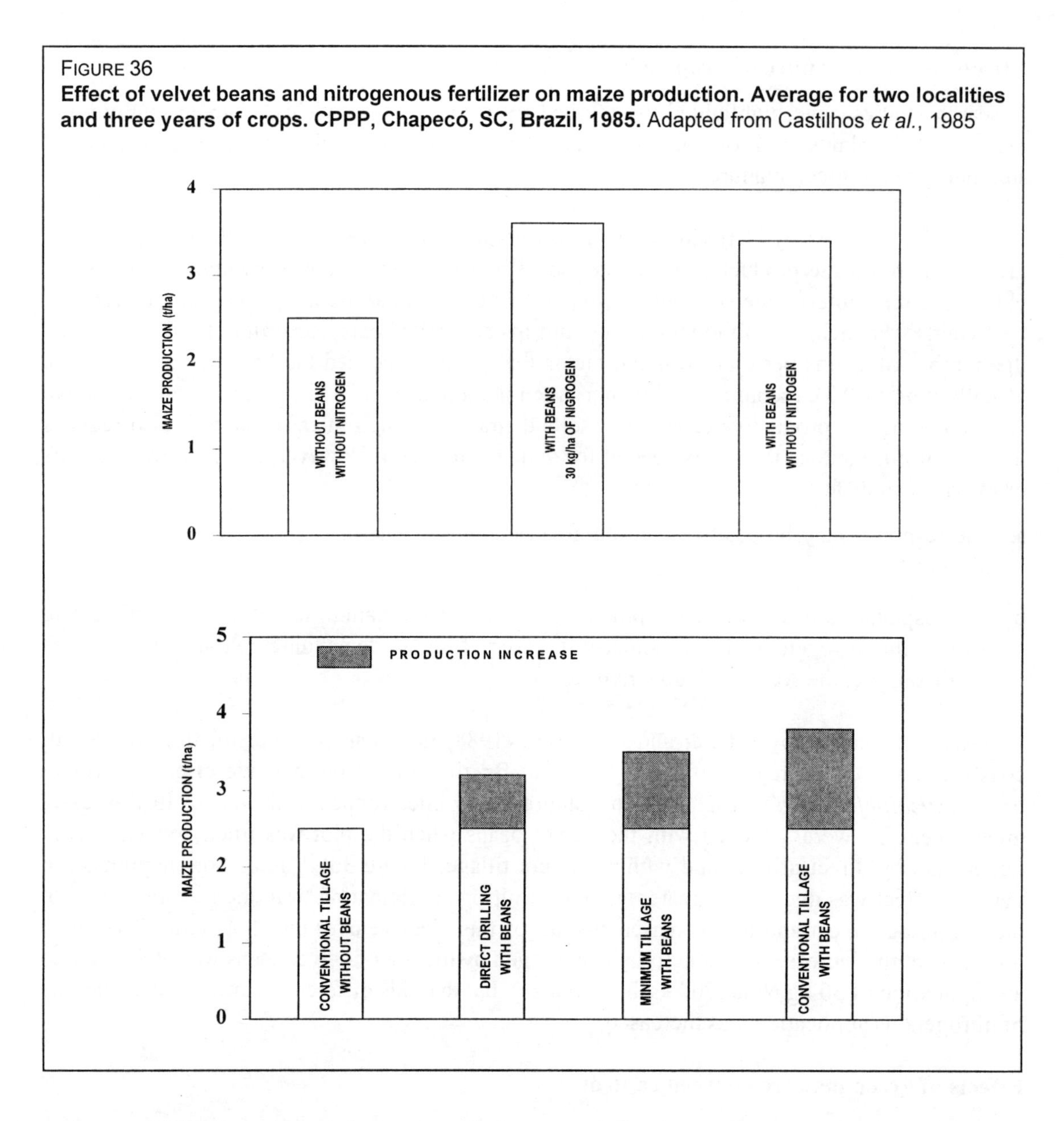

FIGURE 36
Effect of velvet beans and nitrogenous fertilizer on maize production. Average for two localities and three years of crops. CPPP, Chapecó, SC, Brazil, 1985. Adapted from Castilhos *et al.*, 1985

to Lorenzi (1984), amongst the range of plants used for cover, it seems that graminaceous species (maize, wheat, oats, barley, rye) exert the most pronounced allopathic effects, whilst the legumes (lupine, serradela, velvet beans) are also efficient on different species of weed. As regards the relationship between broad leafed and graminaceous weeds, Almeida (1987) observed that in dead cover of lupine, forage turnips and cabbage, the graminaceous weeds dominated the broad leafed weeds, whereas with rye and triticale the domination was reversed (Figure 37). In wheat and oats, there was little domination of one group over the other. The same author also observed that in covers of naked oats, rye, forage turnips and cabbage, it was these that left the field cleanest after the harvest and also those that had a more prolonged effect on the weeds.

Almeida (1985a, b) concluded that by selecting carefully the winter cover crops, it is possible to significantly reduce weed propagation and by this, also reduce the need for chemical herbicides.

FIGURE 37
Influence of dead vegetative cover from various winter crops on the percentage distribution of graminaceous and broad leaf species (Almeida, 1987)

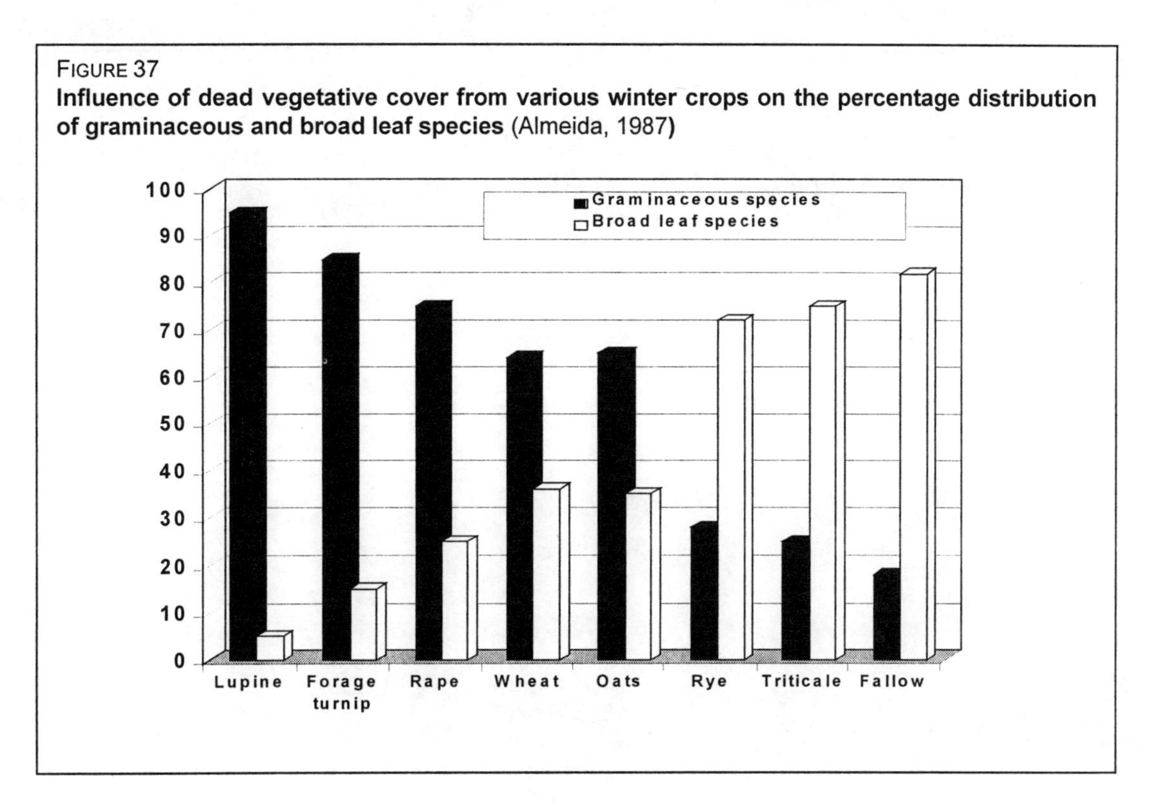

Despite the advantages of using winter crops for forming the dead vegetation cover and of their allopathic effects, one must consider that they also give rise to re-sprouting. Serradella and peas produce very few new shoots but show a problem to naturally re-seed themselves. Generally speaking, the cereals (naked oats, rye, wheat and barley), sunflower, forage cabbage, cabbage and forage turnips show strong re-sprouting, which necessitates the use of herbicides for their elimination. Oats also re-seed naturally (Almeida, 1985a).

Chapter 13

Physical barriers for the control of runoff

Of the total amount of rainfall arriving at the soil surface during intense storms, part infiltrates and the remainder becomes runoff, which concentrates in natural depressions and runs downhill until it reaches natural zones of deposition (level areas and drainage channels). As runoff increases, so does its velocity, volume and its ability to cause erosion. The critical runoff velocity, at which soil particles that have been detached from soil aggregates begin to be transported over the surface, is 5 m/s in sandy soils and 8 m/s in clay soils (Rufino, 1989).

Efficient control of soil erosion due to rainwater can be achieved by the following soil conservation practices:

- Systematic planning and protection of the area from runoff;
- Land preparation;
- Cultivation of crops;
- Soil cover.

The practices involved in the systematic planning and protection of an area against runoff are conceived with the aim of establishing drains or obstacles to channel, retard or retain the runoff. The establishment of these practices will cause changes to the land surface. Traditional practices for achieving these ends include stormwater diversion drains, individual terraces, and permanent vegetation barriers (Sobral Filho *et al.*, 1980).

TERRACING

To ensure that terraces are integrated into local land planning without incurring unnecessary problems or costs, projects for establishing terraces should consider the whole property, a group of neighbouring properties, or even the whole hydrographic micro-catchment. In this respect, the possibility of relocating agricultural areas, fences and roads to ensure the optimum utilization of land should not be excluded.

Study of the area to be terraced

Once it has been decided that terraces are necessary to protect an area, a study should be carried out on the existing land use and nature of the soils. The location of natural drainage lines, low-lying areas, and sites suitable for constructing runoff diversion drains should be recorded.

L. do Prado Wildner
Enterprise for Agricultural Research and Rural Extension of Santa Catarina (EPAGRI)
Small Farm Research Centre (CPPP)
Santa Catarina, Brazil

Where runoff occurs in neighbouring areas, whether along tracks or in gullies, should also be identified. Slope gradients, slope lengths, the presence of rill erosion or gullies, and the location of roads and tracks should be noted, and information on the pattern of rainfall in the area should be collected.

Land levelling

Before any conservation practices to prevent erosion are undertaken, the area may need to be made more uniform by levelling. This can be done using motorized graders, harrows, ploughs, animal traction equipment or even manually. The aim should be to fill in depressions, holes and unnecessary ditches, to level rills caused by erosion, pull out stumps and roots, and might even include eliminating an old terrace system.

Land levelling is often costly but is the only solution for recuperating and then benefiting from the entire agricultural area in situations with pronounced microtopographical variations (Zenker, 1978).

Gully control

Gullies are large erosion features that often devalue agricultural property by hindering the movement of machinery and crop establishment, and by reducing the area that can be farmed. In general, the recovery of gullies is a slow and difficult process and requires a range of practices. These may include closing off the area, constructing stormwater diversion drains on the headland, slowing down flow rates, planting vegetation to protect gully sides, and installing transverse lines of stakes across the gully profile. More details are provided in Chapter 14, *Gully control*.

Construction of stormwater diversion drains

A common mistake when working on terraces or systems of terraces for annual crops is to consider only this area, and to disregard the impact of external influences.

Runoff originating upslope from roads and gullies can damage any terrace system, however well planned. For this reason it is often necessary to construct stormwater diversion drains, which are contour drains with a bank on the downhill side which intercept and divert the runoff into safe waterways, thereby protecting the soil conservation practices in the lower-lying land.

The cross-sectional dimensions of the stormwater diversion drain should be calculated according to the area of higher-lying ground from which the runoff originates, and the quantity of water to be diverted. It is recommended that the waterway be sown with grass to increase its stability and avoid possible problems of gully erosion.

Location of roads and tracks

The planning of soil conservation for a property should take account of the need for roads and internal tracks to be laid out and constructed in such a way that access to the production areas is ensured throughout the year.

Roads should be constructed whilst the terrace system is being laid out, and the main roads should be located along crest lines so that runoff is directed towards the terraces instead of running over the road. Internal paths and tracks designed for the passage of agricultural machinery and implements should be situated immediately below terrace risers, on the side opposite to the terrace channel.

Construction of safe waterways

Safe waterways are natural drainage lines or are specially constructed drainage lines that lead the runoff from stormwater diversion drains and channel terraces downslope to lower-lying areas. They should be protected with native vegetation, and designed with a shape and cross-section capable of conducting the maximum expected runoff without risking erosion of the sides or channel of the waterway. Normally, safe waterways can be established by taking advantage of natural drainage lines, depressions, fields under pasture, or the edges of thickets, woods and bush areas.

Design of terraces

Terraces can be defined as mechanical structures comprising a channel and a bank made of earth or stone. They are systematically constructed perpendicular to the slope. Thus terraces intercept runoff, and encourage it to infiltrate, evaporate or be diverted towards a predetermined and protected safe outlet at a controlled velocity to avoid channel erosion (Figure 38).

FIGURE 38
Schematic representation of the terrace profile showing zones A: of earth movement; B: the bank, and C: the channel. Source: Lombardi Neto *et al.* (1991)

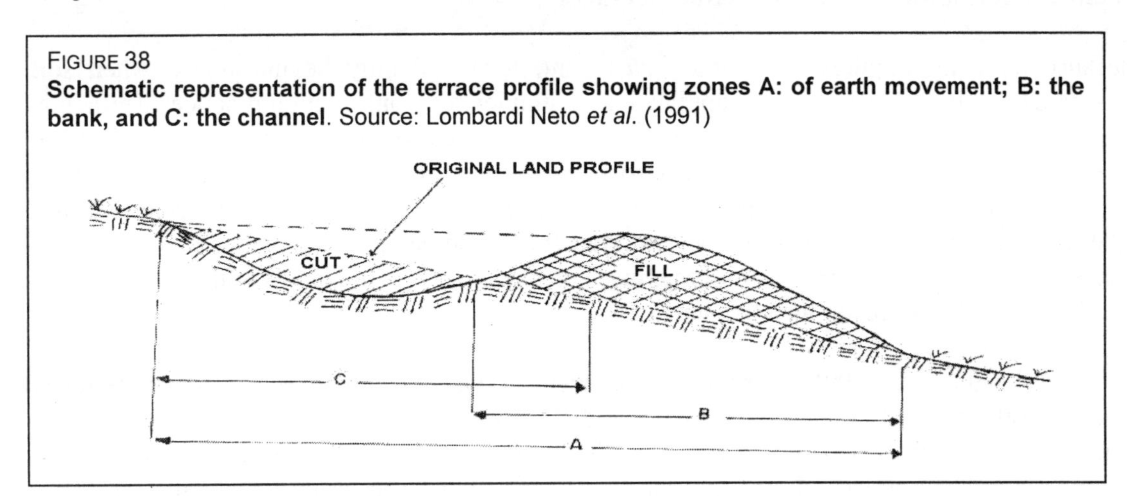

Terraces can considerably reduce soil loss due to erosion if they are well planned, correctly constructed and properly maintained. Results obtained in Paraná (IAPAR, 1984) showed that terracing makes it possible to reduce soil losses by half, quite independently of the system of cultivation employed.

The efficiency of a terrace system will also depend on the adoption of other conservation practices such as contour sowing, strip cropping and soil cover. Other factors to be taken into account are the local conditions, the dimensions and form of construction of the terraces, how well they function and their stability (Rufino, 1989).

Principal objectives of terraces

- To reduce the velocity of runoff;
- To reduce the volume of runoff;
- To reduce the losses of soil, seed and fertilizer;
- To increase soil moisture content through improved infiltration;

- To reduce peak discharge rates of rivers;
- To smooth the topography and improve the conditions for mechanization (Figure 39).

FIGURE 39
Schematic representation of a terrace showing runoff and water movement in relation to slope. Source: Lombardi Neto *et al.* (1991)

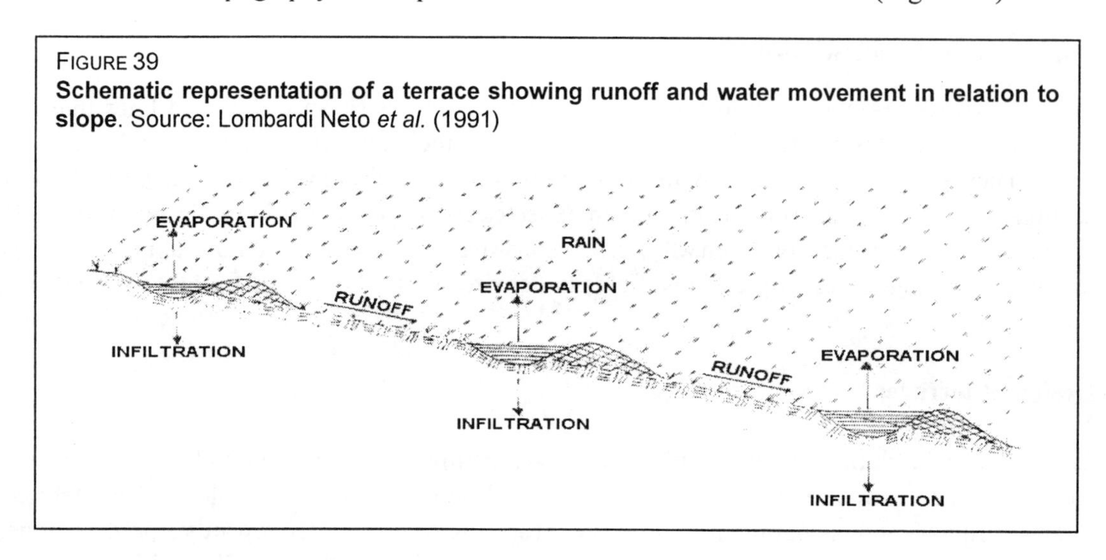

Factors determining when to introduce terraces

Because terracing requires substantial investment, it should only be introduced when soil erosion cannot be controlled by the application of simpler soil conservation practices. Terracing is useful in situations where runoff is common but cannot be adequately controlled by other soil conservation practices, and where the intensity and volume of runoff surpass the water storage capacity of the soil. Terraces are generally recommended for slopes of 4 to 50 percent (Rufino, 1989). Short slopes of less than four percent gradient should be protected by contour vegetation barriers, contour planting or by strip cropping. For longer slopes, the land should be terraced if the slope gradient is greater than 0.5 percent.

Bertol and Congo (1996) proposed new concepts to be considered in the design of terraces when conservation tillage, minimum tillage or direct sowing are employed: 1) the critical slope length, and 2) the presence of crop residues. Using these concepts it is possible to increase the distance between terraces in the presence of soil cover – from direct sowing or other conservation practices – in comparison with the recommended terrace spacing for traditional cropping systems based on ploughing and harrowing.

Classification of terraces

Various criteria are used to classify terraces, of which the most common are:

Destination of the intercepted water

Absorption terraces – these are level terraces designed to accumulate and retain runoff in the terrace channel so that it will eventually infiltrate and the sediment accumulates. These terraces are recommended for low rainfall areas, permeable soils, and for land of less than 8 percent slope. They are normally broad-based terraces.

Graded terraces – these are sloping terraces, designed to intercept runoff and divert the excess that has not infiltrated into protected waterways. These terraces are recommended for high rainfall regions, for slightly or moderately permeable soils, and for slopes of between 8 and 20 percent. They are normally narrow- or medium-based terraces.

According to Bertolini *et al.* (1989), selection of the type of terrace should also take into account those soil physical properties which determine the permeability of water through the profile. Hence, when planning a terrace system it is important to consider the texture, structure, effective depth, and permeability of the surface and subsurface soil.

Construction process

Channel or Nichols terrace – These terraces have more or less a triangular cross section, and are constructed by excavating soil from the upper side only to form a channel, and depositing it downhill to form a bank. They are recommended for slopes up to 20 percent, and are normally constructed with reversible implements, with animal traction or manually. They are used in high rainfall regions, and in soils of low or medium permeability.

Ridge or Mangum terrace – These terraces are constructed by excavating the soil from both sides (downhill and uphill) of the demarcated line to form a bank or ridge with channels both sides of the ridge. They are constructed with either fixed or reversible implements, and are recommended for slopes less than 10 percent, for low rainfall regions, and for permeable soils.

The availability of agricultural machinery and the slope of the land are the main factors that determine which terrace construction method is selected (Bertolini *et al.*, 1989).

Size of the terrace base and distance of soil movement

Narrow-based terraces – where soil movement is limited to about three metres. These terraces include contour ridges.

Medium-based terraces – where soil movement is three to six metres.

Broad-based terraces – where soil is moved more than six metres (but normally less than 12 metres) (Figure 40).

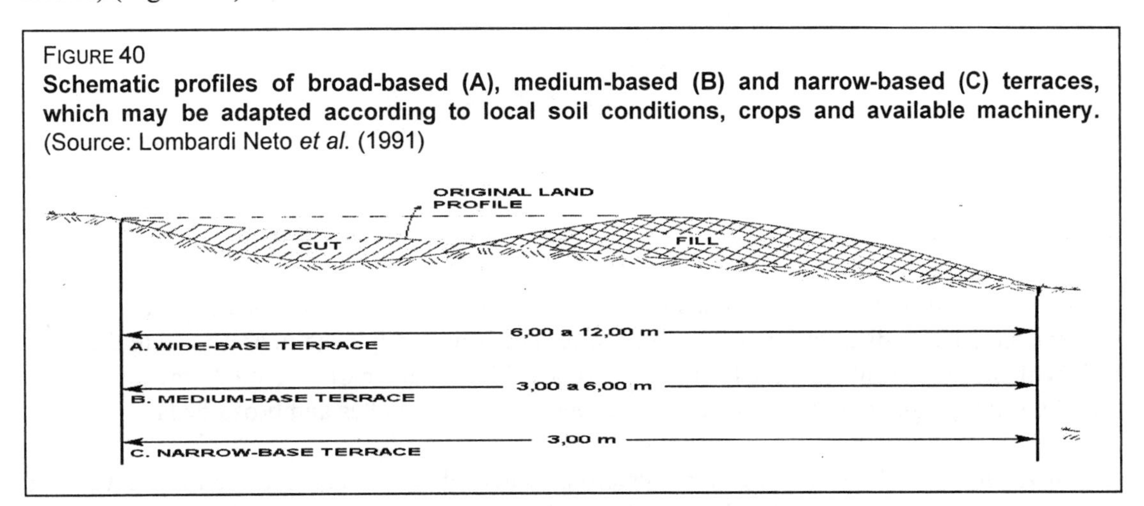

FIGURE 40
Schematic profiles of broad-based (A), medium-based (B) and narrow-based (C) terraces, which may be adapted according to local soil conditions, crops and available machinery.
(Source: Lombardi Neto *et al.* (1991)

Factors which influence the selection of the type of terrace on the basis of the distance of soil movement are land slope, mechanization intensity (tillage and cultivation systems), availability of machinery and equipment, and the farmer's financial resources (Bertolini *et al.*, 1989).

Terrace shape

Another way to classify terraces is according to their shape (Bertolini *et al.*, 1989). In this case, it is slope which is the factor that determines whether a common (Figure 38) or bench terrace should be constructed (Figures 41 and 42).

Figure 41
Schematic profile of an inward-sloping bench terrace with a small gradient towards the riser (b); the inclination of the riser varies according to soil texture. Source: Lombardi Neto *et al.* (1991)

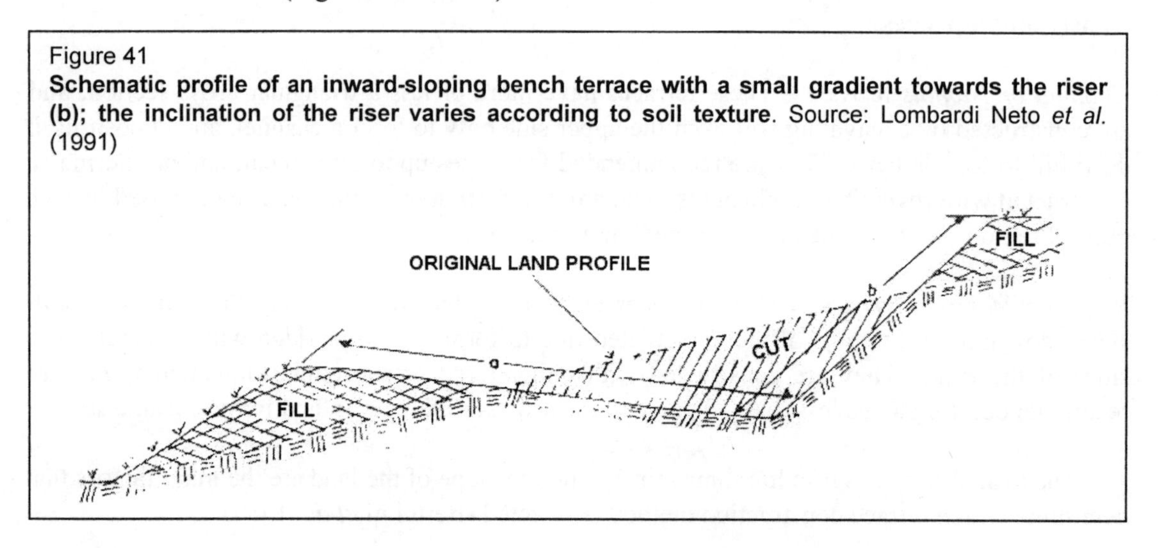

FIGURE 42
An inward-sloping bench terrace changes the land profile into a series of cultivated platforms planted to economic crops without erosion problems. Source: Lombardi Neto *et al.* (1991)

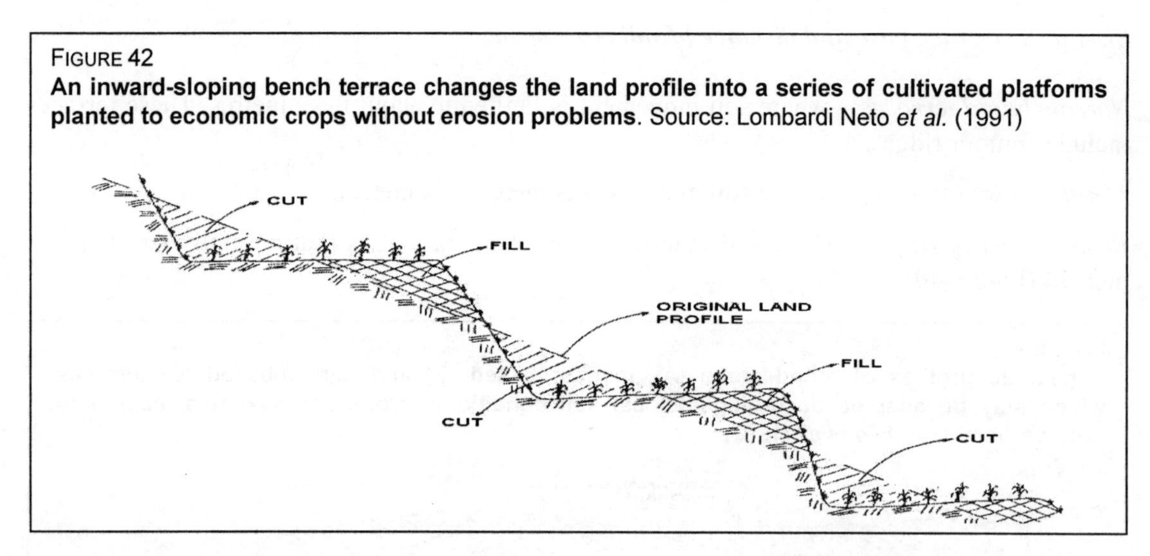

Common (or normal) terraces –consist of a ridge or bank and a channel, which may be constructed on a gradient or level. This type of terrace is normally used in areas where the slope is less than 20 percent. Included within this classification are narrow-, medium- and broad-based terraces, together with some variations such as hillside ditches and broad beds.

Bench terraces – These are true terraces from which all other types of terrace have been derived, and form a series of level or nearly level steps or benches set out along lines of equal contour. They are constructed by cutting and filling, and are used in land with slopes in excess of 20 percent.

Marking out terraces

The planning and construction of a system of terraces requires technical knowledge, which needs to be combined with practical skills and common sense. Every effort must be made to implement a system of terraces that minimizes erosion and encourages rainwater infiltration. To achieve this it is necessary to observe the following important points regarding the location and marking out of terraces:

- Use a level to determine the highest point in the area to be terraced (the field, farm or micro-catchment);
- Identify the line of greatest slope, which is the starting point from which the location of the terraces is decided:
- Determine the gradient of the line of greatest slope using an optical level, clinometer or “A” frame:
  - ❑ If the gradient of the line of greatest slope changes, divide the line into sections of uniform slope.
  - ❑ When there is a change in gradient of the line of greatest slope from one section to another, determine the gradient of the next uniform section, starting from the terrace already marked out in the previous section.
- After calculating the slope and verifying the soil texture, determine the vertical interval (VI) or horizontal spacing (HS) to be used for marking out the terraces:
  - ❑ As a safety measure, locate the first terrace in the highest part of the terrain, at half the recommended distance shown in the table.
- Starting from the stake located on the line of greatest slope, mark out the line of the terrace using stakes every 20 m, or every three steps made with the “A” frame:
  - ❑ In areas that are not very uniform, the distance between stakes should be reduced to 15 or 10 metres.
  - ❑ For absorption or level terraces, all the stakes should be inserted at the same elevation as the stake on the line of greatest slope.
  - ❑ For drainage or graded terraces, the stakes should be inserted at elevations calculated according to the required terrace gradient as specified in Table 33.

TABLE 33
Length, slope and drop in level for graded terraces

Length of the terrace	Gradient	Drop in level	
		Every 10 m	Every 20 m
0 to 100 m	Level	Level	Level
100 to 200 m	1 : 1 000 (0.1%)	1 cm	2 cm
200 to 300 m	2 : 1 000 (0.2%)	2 cm	4 cm
300 to 400 m	3 : 1 000 (0.3%)	3 cm	6 cm
400 to 500 m	4 : 1 000 (0.4%)	4 cm	8 cm
500 to 600 m	5 : 1 000 (0.5%)	5 cm	10 cm

Note: The slope 5:1 000 (0.5%) is the recommended maximum slope limit at which water velocity will not cause erosion in the terrace channel.

- The maximum length of a terrace should not exceed 600 m in clay soils and 500 m in sandy soils:

- In the event that it is necessary to establish a terrace longer than these limits, a new terrace system should be planned sloping in two directions, either towards a central or two lateral drainage canals. Alternatively, the length of the individual sloping sections should be equalised by dividing the terrace into six sloping sections with successive gradients of 0, 0.1, 0.2, 0.3, 0.4 and 0.5%.
- For level terraces, there is no limit to terrace length. However, because of irregularities that may occur in the location or construction of the terraces, it is advisable to construct "cross ridges" every 100 or 200 m to restrict the flow of water.

- When marking out the end of a terrace, increase the slope over the last 20 m to facilitate the discharge of water and to prevent any excess water running down the waterway from flowing into the terrace channel.
- After staking out the line of the terrace, smooth out or realign the stakes to avoid any irregularities, but without changing its slope. This will facilitate the construction and functioning of the terrace.

Terracing and direct drilling

Discussion on whether or not to retain terraces in areas where direct drilling is practised is very controversial.

It should be realized that a significant proportion of rainfall may not manage to infiltrate even under direct drilling, and so may run over the surface as runoff. This will vary depending on soil permeability, degree of soil cover, slope of the land, and the frequency and severity of erosive rainfalls. For this reason, very careful analysis must be carried out before deciding whether or not to eliminate terraces in areas where direct drilling is practised.

SAFE WATERWAYS

A system of graded terraces requires the construction of one or more waterways to receive the water draining from the graded terraces and to safely conduct it to lower parts of the landscape without the risk of erosion.

Classification of waterways

- ***The nature of the channel***

  Natural drainage channels – whenever possible preference should be given to the use of natural waterways with a dense and stable vegetative cover (bushes, thickets or well established pastures) to resist runoff, because of their stability and low cost.

  Artificial drainage channels – when there is no possibility to use natural waterways, artificial channels need to be constructed to receive the drainage water from the terraces.

- ***The shape of the channel***

  Triangular- or "V-" shaped channels – these are recommended for small areas and on gentle slopes as they allow higher flow velocities, impede sediment deposition, and allow flow through small depressions. They can be constructed with blades, ploughs or graders.

Trapezoidal-shaped channels – these are recommended for steeper slopes. Due to their flat base, the water spreads out and so considerably reduces the velocity of flow. The slope of the channel banks should be constructed at an inclination of 4:1, which facilitates construction and maintenance. They can also be made with blades, ploughs or graders.

Parabolic-shaped channels – these are recommended for medium slopes. They best simulate the conditions of natural waterways but are more difficult to design and construct, normally requiring the use of bulldozers.

Construction of waterways

It is essential that the marking out, construction and stabilization of the waterway with vegetation be undertaken before marking out the rest of the terrace system. If this is neglected, serious erosion problems can occur in the waterway, even causing the formation of gullies. For more detailed information concerning the construction of waterways, relevant technical manuals should be consulted.

For stabilizing the waterways, it is recommended that carefully selected plant species are established which should be able to withstand the local temperature variations and long periods of drought. The selected species should not be unduly affected by occasional flooding and should give good soil coverage. They should possess an aggressive root system, well able to hold the soil together, and give plants the firmness to resist the drag forces of flowing water. Finally, they should not be so aggressive that they invade the crops (Sobral Filho *et al.*, 1980). Trials undertaken in Paraná (IAPAR, 1984) led to the recommendation of pangola grass (*Digitaria decumbens*), Bermuda grass (*Cynodon dactylon*) and star grass (*Cynodon plectostachys*). Star grass and pangola grass show rapid growth rates with better soil cover than Bermuda grass. Star grass, because it is highly aggressive, is recommended for lining grassed waterways and drainage channels along roadsides. Pangola grass is more invasive than star grass and can be established during land preparation. All three species are recommended for waterways with steeper slopes and longer lengths. For channels with only shallow slopes, carpet grass (*Axonopus compressus*) is more appropriate.

For the maintenance and care of the waterways:

- Establish the grass cover immediately after marking out and constructing the waterway;
- Establish and protect the vegetation lining the waterway to a distance of at least one metre either side of the banks;
- Use stones or earth banks in the waterway until the vegetation is established;
- The waterways should be wide, sloping downhill and should never be reduced in width as a result of tillage;
- If gullies exist in the waterway, they should be recuperated;
- Never use the waterway as a road;
- Do not discharge waterways in inappropriate places and use diversion drains to direct the water to selected areas;
- Never use gullies, neighbours' properties, roads, thickets or sloping pasture land as waterways;
- Never allow waterways to be used for manoeuvring machinery or for grazing livestock;
- Keep the waterway free of invasive plants;

- Maintain the grass at a minimum height of 10 cm;
- Periodically reseed or replant bare areas in the waterways;
- Carry out periodic conservation work on the waterway channel and banks, particularly at the discharge points of the terraces;
- Ensure that technical specialists carry out the location, construction and maintenance of the waterways (Zenker, 1978 and Daniel, 1989).

Progressive bench terraces with stone or vegetation barriers

Progressive bench terraces with barriers of stones or vegetation strips are conservation practices which, similarly to terraces, are based on the principle of reducing slope length to slow down the velocity and volume of runoff. As sediments are deposited within, and upslope of, the barriers there is a gradual development of bench terraces. Stone barriers are used in stony terrains, and especially in very hilly country. These terraces with stone or vegetation barriers are well adapted to the conditions of small farmers, but should be combined with other conservation practices to achieve an effective control of erosion.

The stone barrier is formed by collecting stones from the surface and placing them in lines across the slope. This practice can be used on land of any slope, but according to Pundek (1985), it is especially recommended for slopes between 26 and 35 percent, and the spacing should be according to the recommendations given in Table 34. A stone wall is then built; the length and height of the wall being determined by the availability of stones and labour.

TABLE 34
Recommended spacing for progressive bench terraces

Land slope (%)	Distance between progressive bench terraces (m)	
	Clay texture	Medium texture
26 to 27	11	10
28 to 29	10	9
30 to 31	9	8
32 to 33	8	7
34 to 35	7	6

Source: Pundek (1985).

The construction of stone barriers requires considerable manual labour and time. For this reason it is normally undertaken with the participation of many farmers, which may awaken a spirit of collaboration within the community.

The advantages of this practice are that erosion is controlled, field operations and improved cropping practices (inputs, spacing, population) are facilitated, and the efficiencies of tillage, seeding and pasture establishment are increased.

In very hilly areas with fewer stones farmers can use strips of vegetation as the barriers, either in isolation or in association with stone barriers.

Vegetation barriers should be made by first marking out a strip (retention strip) of 1.5 to 2 m width, which is then planted with species such as sugar cane, elephant grass, or perennial birdseed grass to stabilize the soil and form a permeable barrier.

After constructing the stone or vegetation barriers, tillage operations should always be carried out with the mouldboard plough turning the soil downslope in the direction of the strip or wall. In this manner, a bench terrace profile is gradually formed.

Contour ridges

The contour ridge is sometimes confused with the narrow-based terrace due to its similar method of construction. They are normally made with animal traction implements or with handtools, although tractor mounted implements can also be used. These structures are more widely used in very hilly regions, on small farms, or in areas which present obstacles to mechanization. Vieira (1987) mentions that contour ridges are particularly appropriate for areas of sandy soils where terracing is not a viable possibility, or for soils with very sandy subsurface horizons. Castro Filho (1989) comments that contour ridges can be used on farms to:

- protect properties from runoff;
- stabilize river banks, irrigation ditches, dams and areas adjacent to bridges;
- recuperate degraded areas where gullies have formed;
- protect road cuttings and steep river banks;
- conserve soil and water in established orchards with reduced inter-row spacing, where the construction of terraces could harm the crop's roots.

Permanent vegetation barriers

Permanent vegetation barriers are strips of vegetation planted along the contour at intervals within the main crop, and consist of perennial species that develop a dense cover capable of reducing the velocity of runoff (Sobral Filho *et al.*, 1980). They are essentially the same as progressive bench terraces with vegetation barriers as discussed above.

Permanent vegetation barriers can be used in annual or perennial crops and, because of their low cost, are an alternative for farmers who do not have the resources to construct terraces. Manual labour, tools or animal traction equipment are needed to construct the vegetation barriers, together with reproductive material of the plants to be sown.

As an isolated practice, vegetation barriers have proved to be efficient in areas with slopes of up to 10 percent (Sobral Filho *et al.*, 1980). For steeper slopes, it is recommended that they are combined with other conservation practices.

The combined use of contour ridges (or absorption furrows) and vegetation barriers have been shown to be a good alternative for controlling surface runoff. As the vegetation barriers are developing, the absorption furrows play an important role in controlling runoff, as well as accumulating moisture for the growth of the vegetation barrier (Castro Filho, 1989).

The combined use of contour ridges and permanent vegetation barriers allows the following:

- Better control of runoff due to the filtering action of the vegetation and the reduction of runoff velocity;
- Due to the filtering action of the vegetation, the accumulation of sediments is encouraged leading to the progressive formation of small bench terraces;
- The possibility of using the vegetative material as livestock feed.

It is recommended that short-cycle species with high root densities and rapid foliage development are used for permanent vegetation barriers.

The species most commonly used are as follows:

- lemon grass (*Cymbopogon citratus*) and vetiver grass (*Vetiveria zizanioides*), rustic species with aggressive root systems, which are not invasive, are easy to propagate and have few flowers. Both grasses also produce aromatic oils used in the perfume industry;
- canary grass (*Phalaris spp.*) which has similar characteristics to the previously mentioned species. It is adapted to colder climates and can be used as pasture for dairy cattle and sheep;
- sugar cane (*Saccharum officinarum*) and elephant grass (*Pennisetum purpureum*). These species produce large amounts of biomass and are used industrially for animal feed and for the production of organic composts. As these are tall crops, they may cause shading and harm the first few rows of the neighbouring or associated crops. These barriers need to be regularly controlled as the cane and elephant grass tend to invade beyond the established strip. An alternative crop is dwarf elephant grass, which has already been tested and approved by small farmers in the southern region of Brazil.

Vieira (1987) also mentions the possibility of using:

- pigeon pea (*Cajanus cajan*), a very rustic species which can also be used for livestock or human feed. It is less efficient than graminaceous crops due to its less spreading root system, but it is a good producer of biomass and an excellent source of proteins for animal feed.
- leucaena (*Leucaena leucocephala*), a legume and excellent forage but more demanding on the soil than pigeon pea. It is also a good source of proteins for animal feed.

Chapter 14

Gully control

Generally, gully control is both difficult and expensive. Restoring an area affected by gullies requires time, labour and capital, for which reason it is recommended to prepare a plan to prevent gully formation.

In addition, land that is badly affected by gullies has an immediately reduced value, justifying measures to contain the problem and at least, to protect the adjoining land and avoid similar consequences beyond the area that is already eroded.

Once gully control has been decided upon, it will be convenient to determine what is the most economic and appropriate protection measure for the area. The cost for correcting the gully and the type of protection to be applied should always be considered in relation to the advantages that may be taken of the investment.

CONCEPT

The so named "gully" represents the most advanced stage of rill erosion. This is the most readily noticeable form of erosion and arises through the surface runoff of the rainwater which concentrates in uneven places or surface depressions that are unprotected or improperly cultivated. Depending on the slope angle and the slope length of the land, the concentrated flow of water in the depressions increases the size of the rills that were initially formed until they eventually become so large as to be called gullies.

GULLY DIMENSIONS

In order to facilitate practical evaluation work in the field, gullies may conveniently be described as follows (Alves, 1978):

- **In relation to their depth**
  - ❑ small gullies, where the depth is less than 2.5 m;
  - ❑ medium gullies, where the depth is between 2.5 and 4.5 m;
  - ❑ large gullies, which are deeper than 4.5 m.

L. do Prado Wildner
Enterprise for Agricultural Research and Rural Extension of Santa Catarina (EPAGRI)-
Small Farm Research Centre (CPPP)
Santa Catarina, Brazil

- **In relation to the catchment area**

  - small gullies, where the catchment area is less than 10 ha;
  - medium gullies, where the catchment area is between 10 and 50 ha;
  - large gullies, fed from a catchment area greater than 50 ha.

MEASURES FOR CONTROL AND STABILIZATION

Although the causes for the deterioration can be completely different, a few basic principles may be applied for a solution to the majority of the cases for recovering or stabilizing the gullies.

Isolation of the gully

The objective of this phase is to stop the process causing the gully, which is the concentration of water that is eroding the bed and destabilizing the sides of the gully.

In order to achieve this objective, it is necessary to establish adequate soil management in the agricultural and other areas that make up the catchment basin (fields, roads, communal areas), so that correct water distribution and infiltration is achieved throughout the basin. Owing to the state of the gully, it is often necessary to construct a terrace or diversion channel immediately above the feed-point of the gully, so as to seal it off completely from further entry of water. On other occasions, depending on the location, the gully needs to be isolated by fencing off the entire perimeter to prevent animals entering or routine fieldwork being undertaken too close to the edges of the gully. It has been found preferable to build the fence at a distance from the edges that is equal to twice the gully depth (Alves, 1978).

Recovery and stabilization of the gully

Depending on the state of the gully and the results of a benefit/cost analysis, one may opt either for the total recovery of the gully or for stabilization of the area, yielding possibilities for its use for alternative purposes.

Recovery

If the size of the gully is not too large and if the expected benefits might be able to compensate for the investment, it is recommended to recover the gully. This entails filling in the gully with earth, recovering the area and incorporating it again within the productive process. This measure is recommended for high-value areas with good productivity of annual crops. Once recovered, the land should be levelled in relation to the adjoining areas and a series of conservation practices implemented to avoid the erosive process repeating itself. The area will have to be levelled again periodically due to the natural settlement of the recovered soil.

Stabilization of the gully

In the event that recovery of the gully is neither technically nor economically viable, it is recommended that the following steps be taken:

- Before closing off the perimeter of the gully, reduce the slope of the banks to prevent them continuing to erode. A tractor with a front-end loader, or manual work using a hoe or a spade can achieve this.

- Once the area has been fenced off, if the growth of natural vegetation is insufficient for good control, some additional plants should be planted or sown, selecting the species according to the future use planned for the area.

- As regards the size and shape of the gully, the following actions are recommended:

  - for small gullies, wider than they are deep, with only slightly sloping sides, or whose catchment area is small, one may use vegetation. In this case, the area could be transformed into a pasture, sowing graminaceous forage. Alternatively, if the area is to be transformed into one of natural protection, a forestry reserve or for industrial production of timber, appropriate trees should be selected, well adapted to the region and showing rapid growth rates. Sowing or planting in lines perpendicular to the slope of the gully is recommended so that they form small defensive steps. Defensive bushes slow down the water velocity inside the gully and cause sediments to be deposited, so encouraging the establishment of fresh vegetation;

  - for larger gullies, it will be necessary to use temporary or permanent structures. Temporary structures should be easy and quick to construct, using cheap materials that are readily available. The most common models of temporary structures are illustrated in the Figures 43 to 48. It is recommended that various structures be constructed along the bed of the gully with heights that do not exceed 40 cm and distributed either at regular intervals or at the most strategic positions. In this way they will encourage the growth of vegetation between the structures. According to the Secretariat of Agriculture for Rio Grande do Sul (1985), temporary structures consisting of stakes can be spaced at intervals similar to those of the terraces and according to the slope of the gully bed (Table 35).

TABLE 35
Construction recommendations for the spacing of staked structures in gullies
(Source: Río Grande do Sul/Secretaría da Agricultura, 1985)

Slope (%)	Spacing (m)
0 to 3	17.0
3 to 6	8.5
6 to 9	5.5
9 to 12	4.1
12 to 15	3.3
15 to 18	2.7
18 to 21	2.3

The barriers or traps made from stakes must be sufficiently well embedded in the bed and the sides of the gully to avoid them being swept away. Attention should also be paid to the design of the weir in the centre of the structure so that it has a section that is sufficient to accommodate the expected volume of flow. As a general rule, the outlet from the barrier should also be protected to avoid the structure being destroyed by the water flow over the drop. It is important to remember that the structures must receive regular maintenance and for this reason, it is recommended that they be inspected for possible damage after heavy storms. This practice is particularly important during the initial phase of installation when the materials are still not consolidated.

FIGURE 43
Barrier made from vegetation (Hull, 1959)

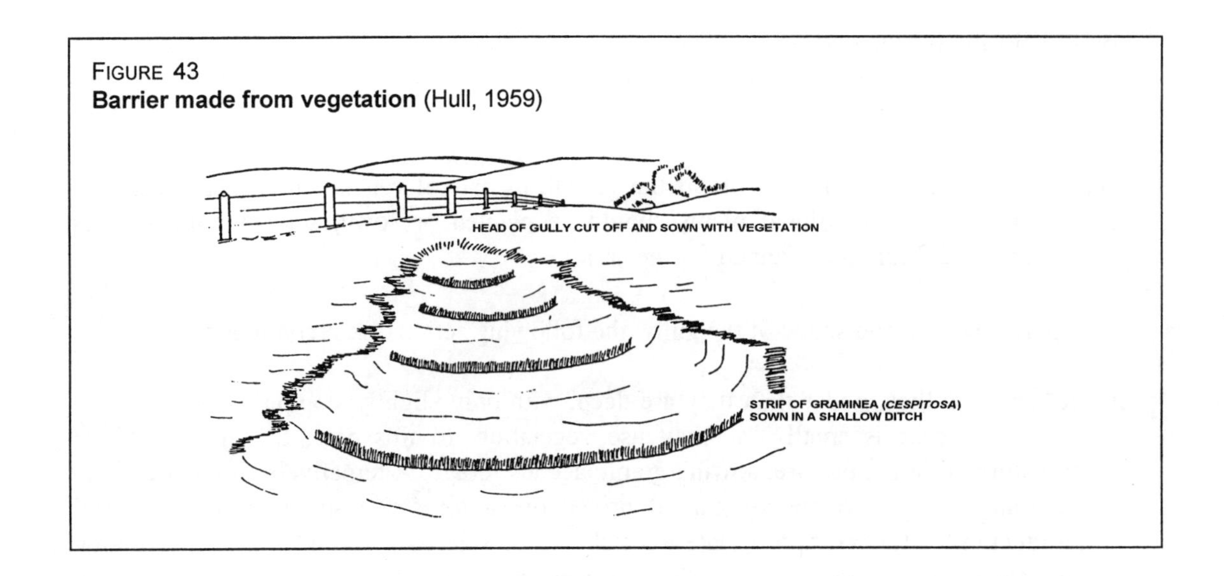

FIGURE 44
Barrier made from branches for normal use (Hull, 1959)

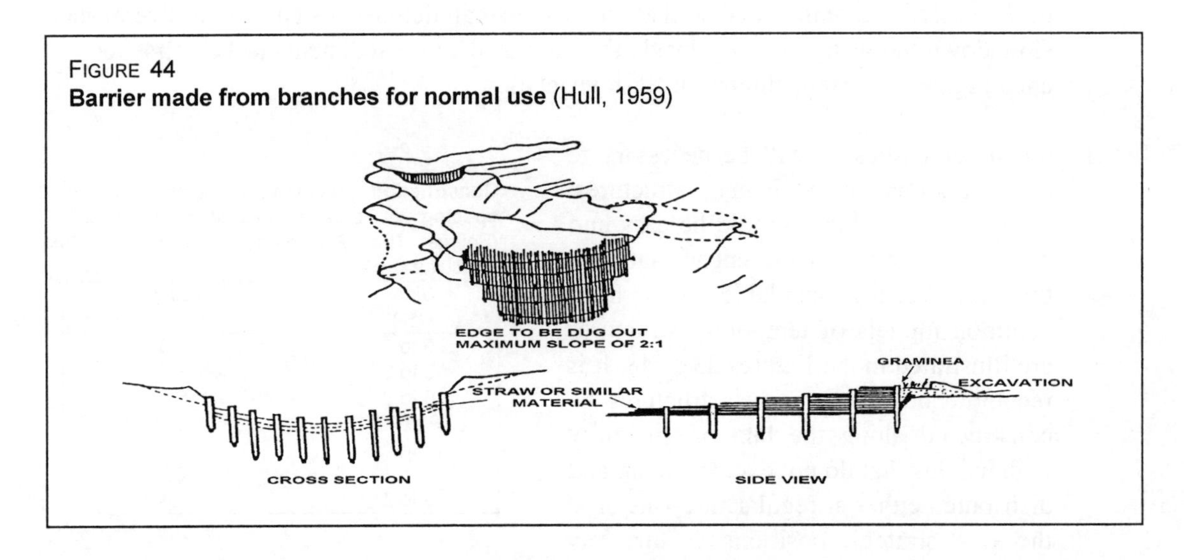

FIGURE 45
Barrier made with branches and wire netting (Hull, 1959)

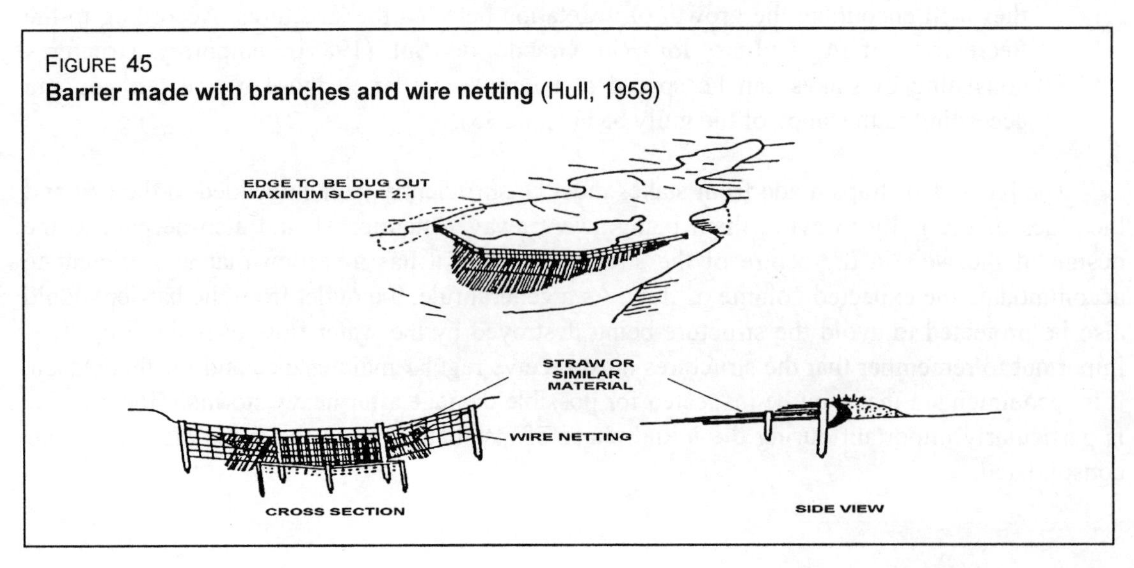

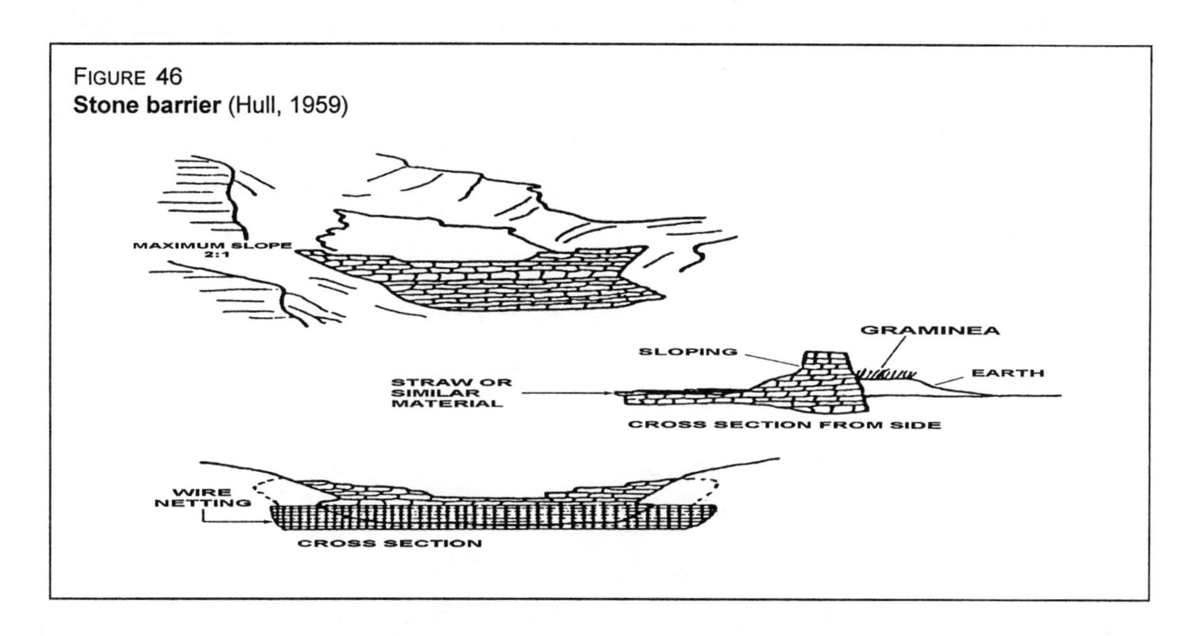

FIGURE 46
Stone barrier (Hull, 1959)

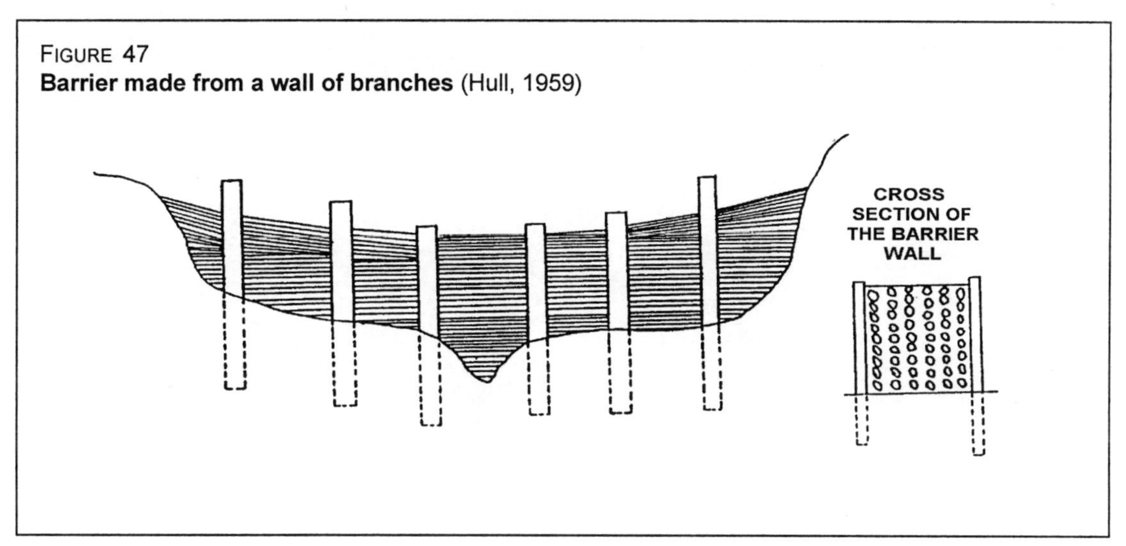

FIGURE 47
Barrier made from a wall of branches (Hull, 1959)

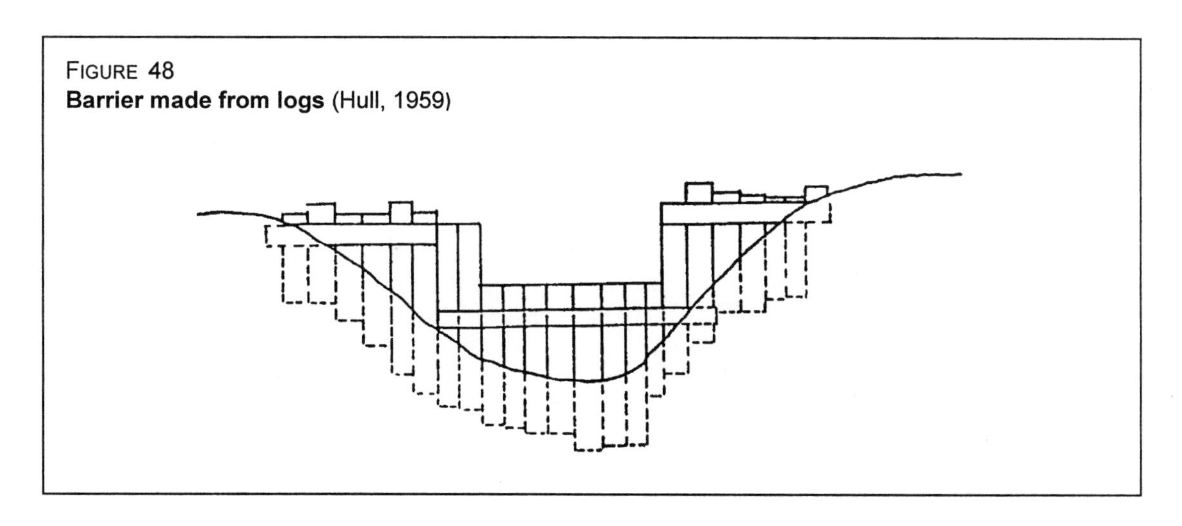

FIGURE 48
Barrier made from logs (Hull, 1959)

Chapter 15

Rainwater capture and irrigation

The semi-arid tropics of Brazil cover an area of 1 150 000 km^2, which corresponds to 70 percent of the region in the Northeast and 13 percent of the country. There is a diverse range of climates, made up of 120 geo-environmental units with large differences as regards their physical, biological and socio-economic characteristics (Silva *et al.* 1993). The limiting factors governing agricultural development in the region comprise, amongst others, a general shortage or poor distribution of the rains, soil limitations and the agricultural practices used.

The soils predominating in the region are of a crystalline origin, normally flat, siliceous and stony, with low infiltration capacity and low organic matter content. Coupled with these characteristics, the high rainfall intensity causes water losses due to runoff and consequent water erosion. Given the characteristics of the region and when planning at farm level, it is convenient to consider minimum risk mechanisms for exploitation which allow satisfactory production, despite the limiting environmental conditions.

Various rainwater harvesting methods using animal traction as the power source have been developed and adopted through the work of EMBRAPA, the Brazilian Enterprise for Agricultural and Livestock Research, based at CPATSA, the Centre for Agricultural and Livestock Research in the Semi-Arid Tropics. The main rainwater harvesting techniques that have proven appropriate for the production conditions of the Brazilian semi-arid zone are presented in this chapter.

PRINCIPAL FACTORS AFFECTING THE ESTABLISHMENT OF SYSTEMS FOR RAINWATER CAPTURE

In order to establish an *in situ* rainwater capture system, information is required concerning a number of factors such as the area to be cultivated, the soil, the topography, the amount and distribution of the rainfall, the crops (annual and perennial) and the availability of equipment and labour. These factors must then be weighed against the socio-economic factors in order to establish the feasibility for investment in the technology.

Rainfall in the municipality of Petrolina, PE, is concentrated in the months from December to April, which is considered to be the agricultural season. Study of the rainfall data over a ten-year

J. Barbosa dos Anjos, L. Teixeira de Lima Brito and M.S. Lopes da Silva
Brazilian Enterprise for Agricultural and Livestock Research (EMBRAPA)
Centre for Agricultural and Livestock Research in the Semi-arid Tropics (CPATSA)
Petrolina, Brazil

period (1985 – 1994) shows a wide variation but in six years out of ten, a concentration of the rainfall in the month of March (Table 36).

TABLE 36
Monthly rainfall recorded at Petrolina, PE, Brazil for the period 1985 – 1994

Month	Year									
	1985	1986	1987	1988	1989	1990	1991	1992	1993	1994
Jan	286.6	36.5	4.8	99.2	27.5	34.2	103.0	344.1	34.4	121.4
Feb	84.9	78.5	30.7	31.7	21.0	90.3	111.2	122.1	12.0	41.0
Mar	172.0	184.1	162.8	256.8	216.7	25.7	207.4	49.4	5.5	156.5
Apr	151.6	25.1	19.8	117.3	154.6	42.8	47.6	33.4	20.8	78.4
May	15.6	14.0	73.3	29.8	78.5	6.6	52.7	4.0	3.8	5.9
Jun	69.9	0.8	0.8	44.9	0.0	9.0	21.8	4.0	5.9	9.9
Jul	5.6	7.7	47.4	5.9	14.1	61.6	0.0	0.9	1.4	14.8
Aug	19.9	1.3	0.0	0.0	5.4	0.3	4.3	0.0	5.8	0.0
Sep	0.0	0.6	0.0	0.0	0.0	0.4	0.0	0.9	0.8	1.0
Oct	3.4	3.6	54.6	14.4	0.0	0.0	0.0	0.0	25.7	0.0
Nov	95.1	3.3	0.0	56.1	6.2	61.3	73.3	31.9	55.5	0.0
Dec	174.0	29.5	12.8	105.0	369.3	18.8	13.1	68.1	16.2	42.2
Total	1 078.6	385.1	407.0	761.1	893.3	351.0	634.8	658.8	187.8	471.1

The sowing date is another factor of extreme importance for the success of dryland agriculture. It has been shown that the best sowing date for cowpea in the municipality of Petrolina, PE, is from March 2 to 6 and for maize, from January 17 to February 9, which coincides with the season of the greatest concentration of rainfall (Silva *et. al,* 1982).

Soil erosion is most affected by the rainfall characteristics, particularly its intensity. In the semi-arid zone of Brazil, the rainfall pattern is characterized by storms of high intensity and short duration, with tillage often being undertaken when moisture conditions are inadequate, causing soil pulverization which consequently makes it more vulnerable to erosion. According to Lopes and Brito (1993), the most critical period for the erosivity of the rainfall is from February to April when almost two-thirds of the total annual erosivity occurs.

Maize cultivation (*Zea mays L.*) requires between 500 and 800 mm of water well distributed throughout the growth phases, without considering other factors affecting its production (FAO, 1979). The stages of flowering and of grain filling are critical for obtaining maximum production.

In conditions with an average rainfall of about 400 mm, it was observed that annually, a moisture deficit occurs due mainly to the irregularity of the rainfall distribution both in timing and the spacing between showers or storms.

METHODS FOR *IN SITU* RAINFALL CAPTURE

The traditional system of minimum tillage in pockets using a hoe, causes a small depression which is capable of storing a certain amount of rainwater. This system is not apparently very aggressive to the environment but, as the soil is not ploughed, the surface appears slightly compacted which makes infiltration difficult and encourages runoff and so contributes to the erosive process. It is therefore necessary to use simple soil preparation techniques for the *in situ* capture of the rainwater, which may be undertaken using either motorized power or animal traction (Duret *et al.*, 1986).

In situ capture of rainwater: ploughing and planting on flat land

Ploughing the soil for establishing dryland crops in the semi-arid north-eastern region of Brazil constitutes one of the *in situ* rainwater capture techniques used in the area. The shaping of small depressions due to the ploughing operation has the objective of impeding surface runoff of the rainwater so that it remains stored in the soil and so available to the crop for a longer period. This system consists of using tractor or animal drawn ploughs, an animal drawn mouldboard plough offering the simplest solution (Anjos, 1985). Figure 49 shows a schematic representation of the crop in the field.

FIGURE 49
Ploughing and planting on flat land

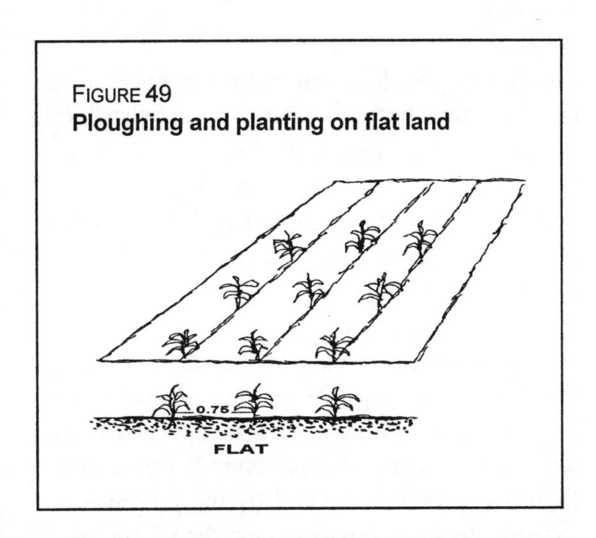

For animal traction, the simplest equipment consists of a mouldboard plough with an 8 inch (0.20 m) working width. It costs approximately US$ 150 and the work animal is valued at around US$ 300, meaning a total investment cost of US$ 450. Hire costs for the same implement and the animal amount to US$ 0.96/h. Using a tractor, the hire costs for a wheeled tractor with a plough vary between US$ 12 and 15/h.

In situ capture of rainwater: ridging after planting

Ridging after planting is a rainwater harvesting technique that consists of ploughing and sowing the flat area followed by ridging between the crop rows and ridging up again a second and third time according to the crop, using either animal drawn or tractor operated ridgers (Figure 50). When crops such as maize and sorghum are well developed, it becomes difficult to use the toolbar equipped with more than a single ridger body. The solution lies in using a single animal to pull a one-row ridger body along the row.

The most appropriate time for ridging cowpea is 20 to 30 days after planting and for maize, 30 to 40 days after planting.

FIGURE 50
Ridging after planting

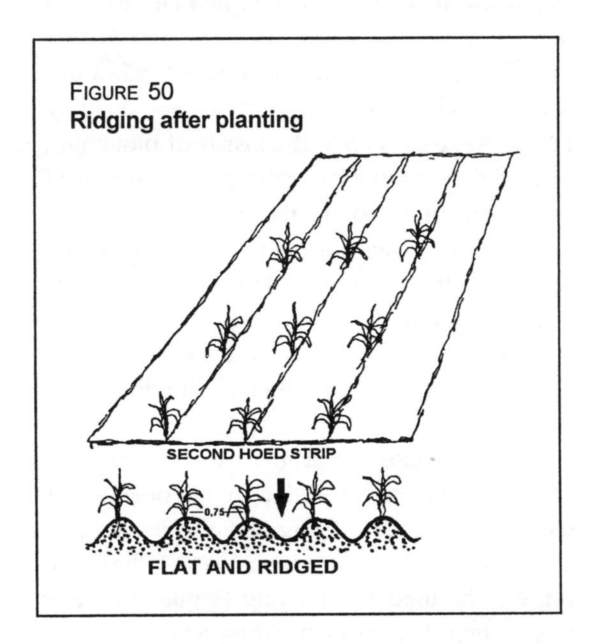

The cost of an animal drawn ridger is approximately US$ 80, which when added to the animal cost of US$ 300, gives a total investment cost of US$ 380. Hire costs for the same implement and the animal amount to US$ 0.90/h. To hire a tractor, the costs for a wheeled tractor with a plough vary between US$ 12 and 15/h.

In situ capture of rainwater: ridging before planting

The technique of *in situ* rainwater harvesting by ridging before planting consists of ploughing the area and then opening up furrows at 0.75 m row spacing. For this system, hoeing is accomplished by ridging along the rows and then using a handhoe to hoe between the plants.

This system, owing to the ridge defining the line of planting (Figure 51), allows better use of the rainwater and also optimizes the weeding, pest and disease control operations. Its use is however, limited by the presence of stumps, stones or slopes steeper than 5 percent.

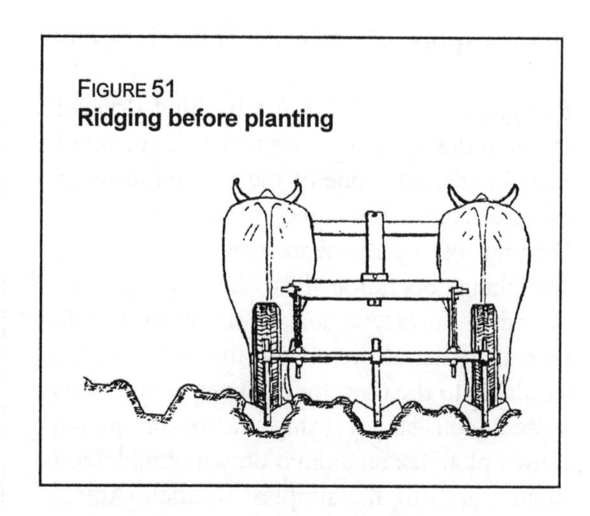

FIGURE 51
Ridging before planting

The cost of an animal-drawn toolbar is approximately US$ 1 500 and for a trained pair of animals, about US$ 1 000. The hourly cost to hire a wheeled tractor is between US$ 12 and 15.

In situ capture of rainwater: tied ridges

The system for *in situ* capture of rainwater using tied ridges has been further developed by EMBRAPA-CPATSA. It consists of ploughing and ridging at a 0.75 m row spacing, followed by an operation to tie the ridges with small mounds along each furrow so as to impede the runoff of the rainwater. Tying the ridges is done with an implement designed for use with animal traction (Figures 52 and 53) and should be undertaken before planting on the ridges.

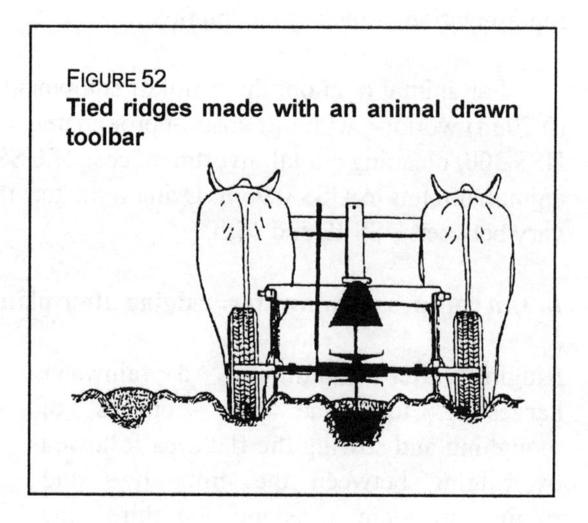

FIGURE 52
Tied ridges made with an animal drawn toolbar

The mounds are at intervals between two and three metres, controlled by the operator of the implement, care being taken to leave them at a height that is less than that of the main ridge to be used for planting (Figure 54). For this system, hoeing or weeding is achieved by using a ridger between the rows and making a second pass with a handhoe between the plants.

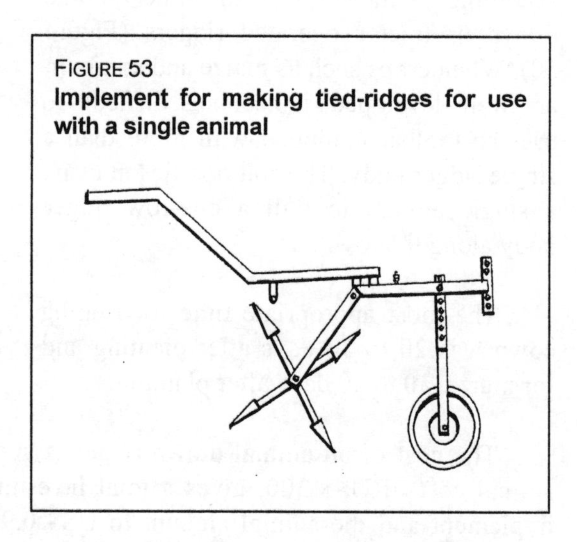

FIGURE 53
Implement for making tied-ridges for use with a single animal

The tied-ridger implement may be constructed in small workshops, by local blacksmiths and costs about US$ 180. It needs little tractive force and can be drawn by small animals such as donkeys, whose average cost is about US$ 70. Hire cost for the animal powered system are about US$ 0.90/h.

FIGURE 54
Tied-ridge system

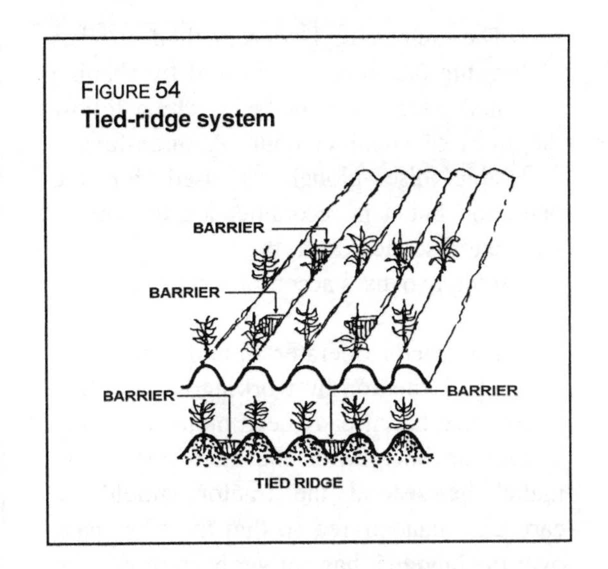

In situ capture of rainwater: partial ploughing

The system for *In situ* capture of rainwater through partial ploughing consists of two successive passes with a reversible animal-drawn plough, leaving a distance of 0.60 m from each second furrows. In this manner, the work time is reduced by half due to the ploughing being accomplished in strips. The unploughed land between the strips is used for harvesting the rainwater, leading it to the seed zone (Figure 55).

FIGURE 55
Strip ploughing system

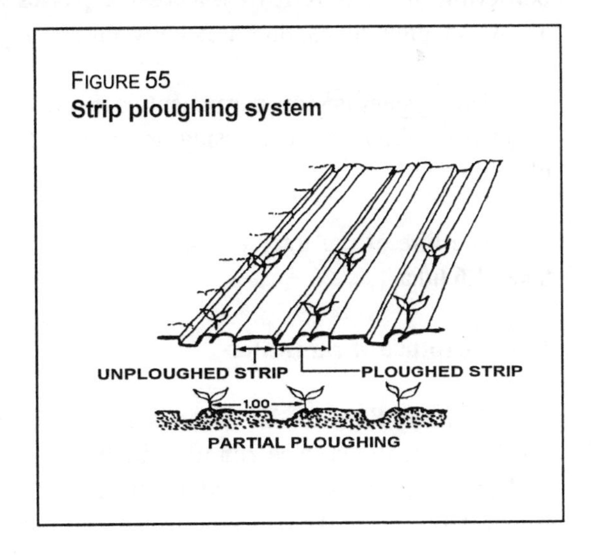

Sowing for this system is accomplished with a punch planter into the second furrow left by the plough in each strip, the inter-row spacing being one metre. The system is re-established, thus promoting a gradual rotation of the cropping area. Hoeing or weeding can be done manually with a handhoe when the plants have reached a height of about 0.10 m. A reversible mouldboard plough may be used, ploughing the unploughed strip towards the plants (ridging) and eliminating the weeds at the same time (Figure 56).

Only a low investment cost is involved in this system. The plough costs US$ 150 and a horse, around US$ 300. The same plough is used for the soil preparation for planting and for the mechanical weeding operation. Hire costs for the implement are US$ 0.70/h.

FIGURE 56
Method for mechanical hoeing or weeding with animal traction

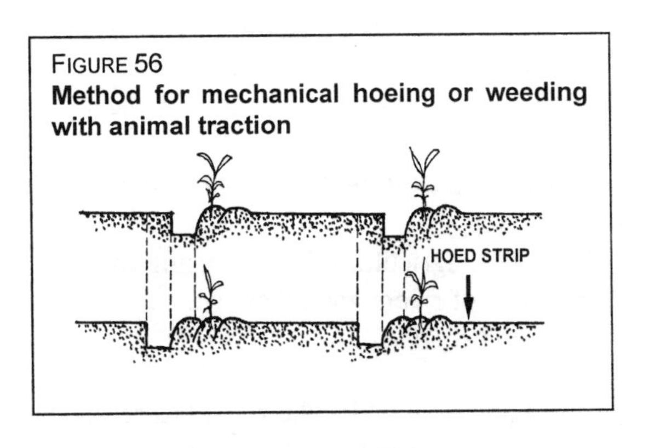

In situ capture of rainwater: the Guimarães Duque method

According to Silva *et al.* (1982), the first *in situ* rainwater harvesting technique, adapted to the semi-arid zone of the north-east, was developed by INFAOL (the North-eastern Institute for the Development of Cotton and Oilseed Crops) and was known by the name "the Guimarães Duque Method for Dryland Tillage". The method was adapted by EMBRAPA-CPATSA for growing annual crops, mainly cowpea and maize.

The Guimarães Duque method consists of forming the furrows followed by shaping high and wide ridges or beds, which follow the lines of equal contour. A three-furrow reversible disc plough is used for the operation but it is recommended to remove the front disc, nearest to the tractor tyres, the remaining two discs accomplishing the task. .

FIGURE 57
The "Guimarães Duque" system of high and wide ridges

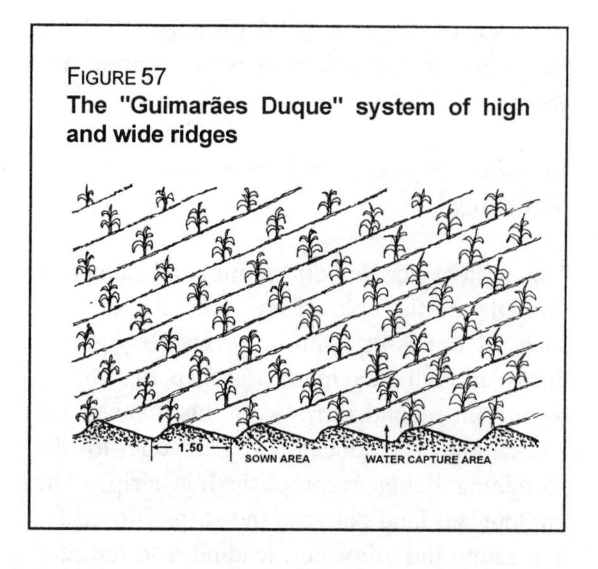

The tractor operator should commence the ridge formation by working parallel to a furrow that has already been ploughed along a contour line. After the first furrow, to make the second the tractor should be carefully manoeuvred so that the tyres pass over the land that has not yet been ploughed, bordering the first furrow. Successive passes are made in a similar manner. This procedure allows shaping the capture area between the ridges, which have a spacing of 1.5 m (Figure 57).

This system is semi-permanent as it can last from three to five years. The system can also be adapted to different crops, using an animal drawn mouldboard plough to loosen the soil in the planting zone.

The hire of the tractor costs between US$ 12 and 15/ha and the ridge formation normally takes 1.6 h/ha.

In situ capture of rainwater

The *in situ* capture of the water is a soil tillage technique, related to the storage of rainwater coming from the surface runoff, which has been studied over the last two decades. It is quite probable that work commenced much earlier during colonial times when plantations of sugar cane were established in the semi-arid North east of Brazil using a system of rectangular pits.

The system is still used today as it provides better moisture conservation in the soil. This is because the soil that has been dug from the pits is then spread around the hole, so breaking up the capillarity and restricting the loss of water.

Cultivating land as the water recedes (recession agriculture)

Recession agriculture implies using the soils that have potential for cultivation in dams, along the sides of rivers and lakes, which are inundated water during the rainy season (Duque, 1973 and Guerra, 1975).

Land from which the water has receded is mainly exploited by small farmers usinghandtools and, to a lesser extent, animal traction. The most commonly grown crops are rice, beans, sweet potato and maize (Carvalho, 1973).

According to Silva and Porto (1982), there are more than 70 000 public and private dams distributed throughout north-eastern Brazil. This allows the survival of some 3 million persons, even during periods of acute drought, exploiting the land as the waters recede (Guerra, 1975).

The most commonly used implement is an 8 inch (0.20 m) animal drawn mouldboard plough. It costs about US$ 150 and the work animal, some US$ 300 giving a total investment cost of US$ 450. The hire cost for the same equipment is some US$ 0.96 per hour. Hire costs for a tractor are between US$ 12 and 15 per hour but it is difficult to work in wet soils as the weight of the equipment makes movement of the machinery very difficult.

IRRIGATION ASPECTS

It is interesting to observe that ancient civilizations had their origins in arid regions, where production was only possible through the use of irrigation. The large populations that were established more than 4 000 years ago along the fertile banks of the Huang Ho and Yangtze rivers in China, along the Nile in Egypt, the Tigris and Euphrates in Mesopotamia and the Indus in the present-day Pakistan, were all conserved due to the use of water resources (Daker, 1988).

The main source of water in north-eastern Brazil is rainwater. Another considerable potential source of water for farming are the surface waters of perennial rivers of which the San Francisco river is the most important, together with water stored in dams constructed in rivers of more normal flow rates. An important, but little exploited resource are the underground waters originating from rainwater or river water.

The capture, lifting and distribution of water for irrigated agriculture is carried out by a number of methods. In the perennial rivers, water wheels are used which take advantage of the hydraulic energy. Small pumps powered by wind are also used. Electric motors or internal combustion engines running on petroleum derivatives or biomass extracts (alcohol, biogas and gasified vegetative matter) power other pumps.

An excess of irrigation water has negative aspects. These may be summarized as the leaching out of soluble nutrients, the high costs for raising the water and also the fact that it can give rise to poor drainage and consequent salinity problems.

The irrational use of water in north-eastern Brazil has caused a rise in the water table. This approaches the soil surface in certain places and during certain seasons, causing unfavourable conditions for crop development, limiting crop productivity and affecting its quality.

Water containing more than 3 g/l of soluble salts is considered poorly suited or unsuitable for agricultural use. Another basic parameter for classifying irrigation water is sodium, for which a content above 0.3 g/l of Na is considered harmful (Valdiviezo Salazar and Cordeiro, 1985).

Irrigation methods

Selection of a particular irrigation method depends on the availability of financial resources, water quality, infiltration rates, soil type, land topography, amongst other factors.

Flood irrigation: this system is characterized by applying a temporary or continuous layer of water to the soil, so totally covering the surface of the land (Soares, 1988).

Furrow irrigation: the irrigation infrastructure is based on engineering works (distribution canals) but it is normally constructed using handtools with soil from the site. The irrigation is not controlled, the water being diverted with earth blockages made with a spade. It would be preferable to use siphons of an appropriate diameter or some other type of flow control system, taking an example from areas being farmed according to adequate planning methods.

Porous capsules: these are receptacles with a capacity of between 0.6 and 0.7 litres connected to a hydraulic water supply network which makes up the irrigation system. They are made from non-swelling clay, which is injected in the form of an aqueous paste into a plaster (gypsum) mould. After removal from the mould and trimming off the edges, an opening is made where a plastic tube will be introduced to carry the water supplied through the network. The capsule is dried and baked to 1 120°C to strengthen it and to attain a porosity of about 20 percent (Silva *et al.*, 1981). The porous capsule irrigation method works at low pressure and has a low water consumption of about 5 litres per unit/day.

Irrigation with earthen pots: these are receptacles with a capacity of between 10 and 12 litres, which are made from clay, dried and baked to give them strength and porosity. Normally, the pots are interconnected with ½ inch (12.7 mm) diameter plastic piping, receiving the water from the supply source. In this way, the labour requirement to supply individual pots is reduced (Silva *et al.*, 1982).

The operational principle of the capsules and the earthen pots is based on when the plants take up moisture from the soil, they generate a difference of water potential between the soil and the porous unit. This causes the water to flow to the soil and to adequately supply the moisture requirements of the crop.

Localized irrigation by drippers and micro-sprinklers: this is characterized by the application of water to the part of the soil that is explored by the plant roots. The application may be periodic or continuous, generally with a pressurized distribution system through small filters and with short irrigation intervals, which maintain ideal moisture levels for the crop (Bernardo, 1982). The operating principles of set localized irrigation systems are illustrated in Figure 58.

FIGURE 58
Schematic diagram of the drip irrigation system

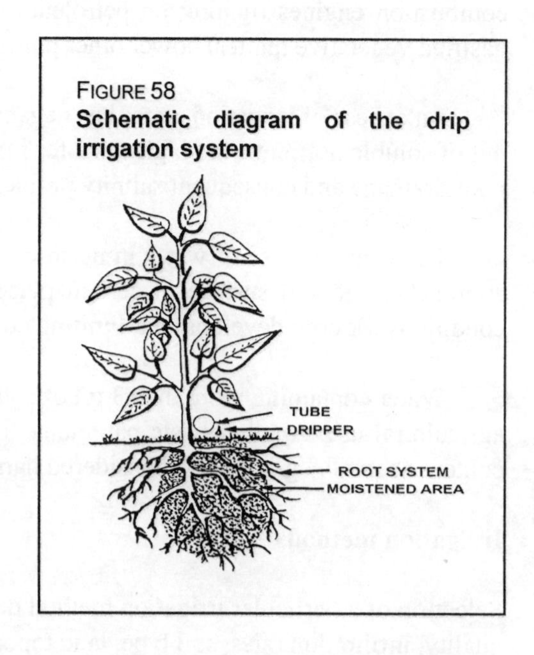

Sprinkler irrigation: this is one of the most widely used systems in recent times. It is characterized by the uniform water application, the high efficiency of the system, the ease to eliminate erosion risks and the possibility for its use in a wide range of different topographies and soil types.

Central pivot: this is a type of sprinkler irrigation system where the unit consists of a lateral line with sprinklers which moves in a circle around a central pivot at a constant and pre-selected speed. Because the system drives itself around the pivot, operational labour requirements are reduced and the system can also directly apply fertilizers and pesticides via the irrigation water (EMBRAPA, 1988).

Soil/water/plant management

Management of the soil water is directly related to the crop planted and to the irrigation system adopted.

Precise definition of the details of the management system may be based on a measure of any of the soil-plant-atmosphere components. When a standard Class A evaporation tank is used, the irrigation schedule may be calculated on the basis of the daily evaporation rate according to the following calculations:

- Calculation of the mean daily evaporation (Ev)

$$Ev = \frac{Ev_1 + Ev_2 + ... + Ev_7}{7} \qquad \text{(Equation 1)}$$

where:
Ev = Mean daily evaporation (mm)
Ev_{1-7} = Daily evaporation (mm)

- Calculation of the daily depth of irrigation required (L_b)

$$L_b = \frac{Kp \times Kc \times Ev}{c_u/100} \qquad \text{(Equation 2)}$$

where:
L_b = Lamina of irrigation water applied (mm)
K_p = Tank Factor, equal to 0.75
K_c = Crop Coefficient
c_u = Coefficient of uniformity of the irrigation system (%), which must be determined locally.

- For drip irrigation and systems with micro-sprinklers, the volume of water to be applied through each application unit depends on the lamina of irrigation water required and the number of plants for each irrigation sub-unit.

  It follows that:

$$V_{ap} = \frac{L_b \times E_p \times E_f}{D} \qquad \text{(Equation 3)}$$

where:
V_{ap} = Volume of water applied per plant (l/plant/day)
E_p = Inter-plant spacing (m)
E_f = Plant row spacing (m)

D = Number of days interval between irrigation applications

- The time required to apply the irrigation lamina will be:

$$T_j = \frac{V_{ap}}{N \times q_e} \qquad \text{(Equation 4)}$$

where:
T_j = Irrigation time for each irrigation unit (hours)
N = Number of drippers per plant
q_e = Flow rate of the drippers (t/h) (this parameter should be determined during field tests)

When irrigation times exceed three hours, it is recommended that the application should be divided into two stages so as to avoid excessive water losses due to deep percolation and asphyxiation of the root system.

- In the case of semi-automatic drip and micro-sprinkler systems, the volume of water per unit should be determined.

$$V = 10 \times L_b \times A \qquad \text{(Equation 5)}$$

where:
V = Volume of water per irrigation unit (m^3)
A = Application area of the irrigation unit (ha)

- For sprinkler irrigation systems, the amount of irrigation water required during the period of greatest development for crops such as tomato, onion, melon and watermelon, may be calculated on the basis of the evaporation accumulated over weekly periods in a Class A tank (Azevedo *et al.* 1986). The calculation uses Equation 6 below:

The amount of irrigation water required is:

$$L_b = \frac{K_p \times K_c \times E_v}{E_I} \qquad \text{(Equation 6)}$$

where:
L_b = Lamina of irrigation water applied (mm)
K_p = Tank Coefficient for a Class A tank (taken as 0.75 or a table established for the particular area should be used)
K_c = Crop Coefficient
E_v = Mean daily evaporation from the tank (mm)
E_i = Efficiency of the irrigation system, determined during field trials

When the soil moisture content is not measured, the soil water availability should be estimated on the basis of tables developed and adapted to the particular locality, so as to determine the time required for the next irrigation operation (Table 37).

TABLE 37

Values of the Tank Coefficient for a Class A tank (K_p) for estimated values of the reference rate of evapo-transpriration (Eto)

	Exposure A Tank surrounded by grass				Exposure B Tank surrounded by bare soil			
UR % (mean)		Low <40%	Medium 40-70%	High >70%		Low <40%	Medium 40-70%	High >70%
Wind (km/day)	Tank position R(m)*				Tank position R(m)*			
	0	0.55	0.65	0.75	0	0.70	0.80	0.85
Slight	10	0.65	0.75	0.85	10	0.60	0.70	0.80
<175	100	0.70	0.80	0.85	100	0.55	0.65	0.75
	1 000	0.75	0.85	0.85	1 000	0.50	0.60	0.70
	0	0.50	0.60	0.65	0	0.65	0.75	0.80
Moderate	10	0.60	0.70	0.75	10	0.55	0.65	0.70
175 – 425	100	0.65	0.75	0.80	100	0.50	0.60	0.65
	1 000	0.70	0.80	0.80	1 000	0.45	0.55	0.60
	0	0.45	0.50	0.60	0	0.60	0.65	0.70
Strong	10	0.55	0.60	0.65	10	0.50	0.55	0.75
425 – 700	100	0.60	0.65	0.75	100	0.45	0.50	0.60
	1 000	0.65	0.70	0.75	1 000	0.40	0.45	0.55
	0	0.40	0.45	0.50	0	0.50	0.60	0.65
Very strong	10	0.45	0.55	0.60	10	0.45	0.50	0.55
>700	100	0.50	0.60	0.65	100	0.40	0.45	0.50
	1 000	0.55	0.60	0.65	1 000	0.35	0.40	0.45

Observations:

For extensive areas of bare soil and in conditions of high temperature and strong wind, reduce the value of K_p by 20%. In moderate conditions of temperature, wind and humidity, reduce the value of K_p by between 5 an 10%.

* R(m) is the least distance (expressed in metres) from the centre of the tank to the limit of the border (of grass or bare soil).

Source: Tuler *et al.* (1983)

- The irrigation time is calculated as:

$$T_i = \frac{L_b}{I_a} \qquad \text{(Equation 7)}$$

where:

T_i = Irrigation time (hours)

I_a = Application rate of the sprinkler, measured in the field (mm/h)

Equation (6) may also be used for furrow irrigation to calculate the amount of irrigation water required (L_b). The irrigation time (T_i) is a function of this amount and of the time for the water to advance along the furrow (T_a), related to the time of opportunity for irrigation (T_o). The times (T_a) and (T_o) are determined directly by field tests.

The use of tensiometers can assist in controlling irrigation, particularly for areas under drip or micro-sprinkler irrigation. They are well adapted for soils where most of the available water is retained at tensions less than 0.80 bar (Faria and Costa, 1987).

Chapter 16

Selection of alternative technologies

INFORMATION SOURCES CONCERNING ALTERNATIVE TECHNOLOGIES

There are various information sources on new soil management technologies. It is recommended to seek published information on technologies that have been investigated in the same recommendation domain where the work is being undertaken, and which have been evaluated and validated by the farmers in their own properties. In this manner, one may have confidence in the suitability of these technologies for the conditions under which they were evaluated.

Often, information is available concerning technologies that are still at an experimental stage and which have not yet been validated by farmers under their own particular conditions. It is not possible to recommend these technologies without further evidence regarding their suitability for the particular zone and conditions of the farmers. When, however, it appears likely that these technologies will be suitable, a programme of participatory research could be organized with a number of farmers interested in the techniques, instead of waiting until the whole sequence of investigations and validations has been completed.

Technicians often possess information from the literature, or have picked up information during visits to other regions or countries, concerning technologies that work well under similar conditions but which have not yet been validated in the country or region where the technician is working. In this case it is very important to compare the agroecological and socio-economic conditions of the farmers where the technology has proved successful with the local conditions so as to assess the likelihood of a successful transfer of the technology. If the probability of success is good, it is suggested that the first step would be to carry out some participatory trials with one or two farmers. If the results are encouraging, the number of participatory trials could be increased so as to obtain more complete information concerning the suitability of the technology, and that the farmers themselves evaluate its appropriateness for their conditions.

The farmers would test the new technology on a small part of the farm and compare it with their normal practice. The tests would be made without plot replications but if possible, the total number of tests should be sufficient to permit statistical analysis of the results as if the tests on different farmers' fields were true replications.

TECHNOLOGY SELECTION ON THE BASIS OF FARMERS' CIRCUMSTANCES

It is very important that any new technologies introduced are consistent with farmers' resources and circumstances. The new technologies should fall within the general framework of the actual

R. Barber, Consultant
Food and Agriculture Organization of the United Nations (FAO)
Rome, Italy

production systems of the farmers. A production system is composed of various components such as land preparation, tillage, crops and their rotation, seeding, possibly application of herbicides and pesticides, fertilization, harvest, grain storage, residue management, machinery, implements and equipment, systems for drainage, irrigation and soil conservation. If a new technology requires changes to some of these components, this will often entail changes to other components. For instance, crop rotation systems become more important when zero tillage systems are introduced as the crop residues remain on the soil surface. This may stimulate the survival and proliferation of certain diseases and pests in the soil that would not occur in a conventional tillage system.

It should not be expected that from one day to the next, the farmer will completely change the production and soil management systems which he or she has practised for many years. The process of change should be gradual, in stages, and at a rate that is acceptable to the farmer.

Socio-economic factors

Both soil management and production systems are influenced by socio-economic factors. It follows that the socio-economic limitations should be taken into account when selecting soil management systems so that they are specially and economically acceptable. The socio-economic limitations that should be considered include the following:

- ***Farm size and production levels***

  The farm size and the level of production influence the profitability and so have a great influence over the most appropriate level of technology to be introduced. Some mechanized technologies, such as zero tillage, require large initial investments in machinery and so are not feasible for farmers with low levels of income. The scale of production can also prove to be a limiting factor for introducing certain management systems with high installation costs such as irrigation and drainage systems.

- ***Financial resources, prices, costs and availability of inputs and credit***

  All these factors influence the profitability of the existing management systems and the possibilities to successfully introduce new production systems. Often, systems which appear technically appropriate are not feasible due to the shortage of economic resources of the farmer, cash flow problems, the high cost of inputs, the difficulty to obtain inputs on time, or low or uncertain crop prices. Many farmers are unable to obtain loans, as they do not have a guarantee that is acceptable to the lending institution. Alternatively, interest rates may be too high or the payback period may be too short, meaning that the farmer would have to sell his entire crop when the market is saturated and prices are very low. In this way these factors can create difficulties for the introduction of new soil management systems.

- ***Marketing, access, transport and crop storage***

  These factors are very important for decisions regarding the introduction of new production systems. It is only worthwhile for a farmer to sow a certain crop if there is a market for it, if transport to reach the market is available, reliable and at an acceptable cost, and if the access roads to the farm are usable. For those farmers living in isolated places where the roads are in a poor state during the harvest season, it would not be advisable to introduce perishable crops. The feasibility to introduce new production and soil management systems may also be influenced by whether or not the farmer has possibilities to store the crop.

- ***Manual labour***

  The availability, cost and opportunity cost of manual labour will influence the selection and profitability of production and soil management systems. There is often a shortage of manual labour at critical times such as during planting, the first weeding operation and harvest, which influences the feasibility of introducing a new production system. For example, cotton production requires considerable manual labour during weeding and harvesting and so is only feasible in zones where sufficient manual labour is available. For small farmers, the availability of manual labour can vary considerably from one family to another, depending on the number and gender of the children.

- ***Farmer organizations***

  The organization of farmers is one of the most important steps in agricultural development. It is more feasible to introduce profitable production systems in situations where farmers are willing to from an organization. This is because they will then be able to purchase the necessary inputs and sell their crops in a co-operative manner at better prices, and will not remain at the mercy of intermediary traders. For the development of irrigation systems it is necessary to organize user groups, particularly to control water use.

- ***Land tenure***

  Land tenure problems are among the most difficult to overcome. They have considerable influence on the soil management systems. If the land upon which the crops are grown is rented, a farmer would be unwilling to contemplate changes or technologies that offer benefits over a period longer than the rental period. Frequently, this leads to annual production systems that are not sustainable because many technologies which promote sustainability, such as soil conservation practices, only yield benefits over the long term.

- ***Technical assistance***

  Not only technical assistance but also frequent and regular stimulation through visits by extension workers is necessary to raise the living standards of farmers. Generally, the farmers that develop most rapidly and successfully are those that receive the greatest direct and personal contact from extension workers. The successful development of new agricultural practices requires the presence of government or non-governmental institutions that are skilful in techniques of technology transfer. In time, selected farmers may be trained to transfer the new technologies to other farmers. However, the first step is the presence of a sufficient number of extension workers in the zone to attend adequately to the needs of the farmers. If these conditions are not present, it will be much more difficult to bring about adoption of new or more advanced technologies by the farmers.

TECHNOLOGY SELECTION ON THE BASIS OF ENVIRONMENTAL CONSIDERATIONS

Soil management systems must not cause negative impacts on the environment, which includes human and animal life, land, water and the atmosphere. The use of toxic pesticides that seriously affect health, even to the extent of causing mortality amongst humans and livestock and which could contaminate the soil causing a reduction in the population, diversity and activity of soil fauna and micro-organisms should be avoided.

In addition, one should avoid management systems that lead to the contamination of waters through leaching of pesticides and nutrients such as nitrates due to excessive applications of nitrogen, or through runoff carrying fertilizers, pesticides and soil particles into the rivers. Water contamination deteriorates the quality of drinking water supplies both for human and animal consumption, damages river and marine life, can affect fisheries and tourist industries, cause sedimentation in dams, and reduce the working life of hydroelectric stations.

It is important to consider the impact of soil management systems not only n the agricultural region where they are applied, but also in downstream areas towards the sea. Sometimes the application of a technology has no adverse effects within the farms, but the sum of the effects on downstream waters can be very harmful. For example, the loss of nitrates due to the excessive application of nitrogenous fertilizer in agricultural areas can cause toxic concentrations of nitrates in the drinking water supplies of downstream populations. In a similar manner, the degree of erosion due to a particular management system could be acceptable to the local farmers. However, the accumulation of sediment in the rivers and the increased turbidity of the seawater could have a negative effect on the growth of corals with a consequent negative impact on the tourist industry. It is therefore fundamental to consider environmental effects in the broadest possible sense.

TECHNOLOGY SELECTION ON THE BASIS OF "PROBLEM – SOLUTION" RELATIONSHIPS

Soil management systems should be developed from the production systems presently used by the farmers. These depend upon the characteristics of the soil, the climate, the land use potentials and very importantly, on the markets, prices, financial resources, and family needs of the farmer. The economic and market factors are very important for farmers who wish to diversify their production so as to generate increased income and improve their living standards. The selection of production systems will not be dealt with in greater detail as these issues are more related to the subject of land evaluation.

In order to select soil management systems, the farmer together with the extension worker must identify the factors that most limit the productivity, profitability and sustainability of the production systems. These include edaphological, climatic, environmental and socio-economic factors. The socio-economic and environmental factors have already been discussed above. In general terms, there is an interaction between some of the edaphological and climatic factors and therefore one considers the limitations imposed by their combined effect, such as the lack or excess of moisture, non-optimum temperatures and strong winds.

This section deals with the edaphological and the combined edaphological and climatic limitations affecting soil management, the causes of these limitations and possible solutions to the problems. A summary of the limitations, their causes and elements of possible solutions is presented in Table 38.

A. Poor crop germination conditions

Poor germination can be attributable to poor seed quality, particularly if the farmer uses his own seed. It may also be due to a lack or excess of soil moisture, to temperatures that are either too high or too low, or to degraded soil structure in the germination zone. Sometimes, the low productivity of subsistence farmers is mainly due to a low plant population.

TABLE 38
Limitations, causes and elements of possible soil management solutions

Limitation	Cause	Elements of possible solutions
A. ***Poor crop germination***	i) Lack of moisture	Leave crop residues Sow a cover crop Conservation tillage; zero tillage Application of mulch Wind breaks Deep seeding for dryland farming
	ii) Excessive moisture	Subsoiling Drainage canals Ridge tillage Cut-off drains Levelling with a levelling harrow
	iii) Excessive temperatures	Leave crop residues Sow a cover crop Application of mulch Conservation tillage; zero tillage
	iv) Low temperatures	Ridge tillage Remove surface residues Wind breaks
	v) large and hard clay aggregates	Tillage with clod-breaking rollers Tillage with a disc harrow or rotary cultivator Fallow the land under pasture crops
B. ***Poor crop emergence***	i) Crusting	Leave crop residues Sow a cover crop Provide a mulch Zero tillage Till into narrow ridges Increase seeding density Reduce seed depth
	ii) Hardsetting layers	Tined tillage with a chisel plough followed by a vibro-cultivator
C. ***Restricted root growth***	i) Severe compaction	Subsoil (for recuperation)
	ii) Incipient compaction	Tined tillage with a chisel plough followed by a vibro-cultivator Tined tillage with controlled traffic
	iii) Hardsetting layers	Tined tillage with a chisel plough followed by a vibro-cultivator
	iv) Excessive moisture	See E ii)
	v) Lack of phosphorus	See D i)
	vi) Presence of toxic substances	See D iv)
D. ***Low fertility and productivity***	i) Nutritional deficiencies or imbalances	Apply fertilizers Apply foliar fertilizers Placement of the fertilizer Split application of the fertilizer Timeliness of fertilizer application Use of non-acidifying fertilizer Incorporation of green manure Crop rotations with legume crops
	ii) Low organic matter and clay contents	Leave crop residues Sow a cover crop Conservation tillage Incorporate green manure Inter-crop with legumes Inoculate the legumes Rotate the crops Fallow with a cover crop Enriched fallows Apply farm-yard manure Apply compost Apply fertilizers Apply lime or dolomitic lime
	iii) Leaching	Rotations with deep-rooting crops Perennial crops Alley crops

Limitation	Cause	Elements of possible solutions
	iv) Aluminium or manganese toxicity	Change the variety or the crop Apply lime or dolomitic lime Apply gypsum with or without lime Incorporate organic manures
	v) Weeds, diseases, pests and exhausted soils	Rotate the crops see D (ii)
E. ***Edaphological and climatic factors*** ***i) Lack of moisture***	i) Evaporation and low infiltration (where residues are available)	Leave crop residues Sow a cover crop Apply a mulch Apply organic manures Zero tillage Change the variety or the crop
	ii) Strong winds	Wind breaks
	iii) Low infiltration rates (where crop residues are not available)	Tied and narrow ridges Strip tillage Tined tillage Field cultivator after each rainfall Plough at the end of the rainy season
	iv) Low moisture retention	Incorporate organic manures Incorporate cover crops Incorporate green manure Subsoil Practice fallowing to accumulate moisture Change variety or crop Sprinkler, gravity or drip irrigation
ii) Excessive moisture	i) Runoff	Cut-off ditches
	ii) High water table or impermeable layers	Open drainage channels Subsoiling Tillage to form ridges Raised beds and cambered beds
iii) Strong winds	i) Lack of protection	Wind breaks
F. *Low biological activity*	i) Shortage of residues	Leave crop residues Apply a mulch Sow cover crops Zero tillage Apply organic manures Sow crops or varieties with more and persistent straw
	ii) "Exhausted" soils	Rotate the crops; see D (ii)
	iii) Toxic pesticides	Apply biological pesticides Apply selective pesticides Use integrated pest management techniques Use integrated weed management techniques
G. *Water erosion*	i) Lack of soil cover and low infiltration rates	Leave crop residues Apply a mulch Apply organic manures Sow cover crops Sow intercrops and relay crops Zero tillage Avoid straw burning Minimize grazing of residues Leave stones on the soil surface Increase plant populations Increase the chemical fertility of the soil Use varieties and species of high biomass Control weeds with herbicides Control weeds with cultivators
	ii) Lack of surface roughness	Contour tillage and seeding Sowing in strips Ploughing after the rains and sowing in strips
	iii) Runoff	Sow live vegetative barriers Alley cropping Stone walls Basins Hillside ditches Diversion drains

Limitation	Cause	Elements of possible solutions
		Bench terraces Orchard terraces Individual terraces
H . ***Wind erosion***	i) Lack of cover ii) Strong winds	Leave crop residues Apply mulch Apply organic manures Conservation tillage Use herbicides Use field cultivators Wind breaks Tillage for making narrow ridges
I. ***High production costs***	i) Labour costs ii) Machinery costs iii) Pesticide and herbicide costs iv) Fertilizer costs	Improved seed drills with fertilizer applicators Zero tillage Modification of the seed drills Systemic herbicides Integrated weed control Organic pesticides Integrated pest management Crop rotations Sow legume crops Use organic manures Apply compost Apply economic rates of fertilizers Apply fertilizers in split doses Place fertilizers Timeliness of fertilizer application Sow cover crops Apply rock phosphate Alley cropping Fallows Enriched fallows
J. ***Environmental Contamination***	i) Toxic pesticides ii) Loss of soluble fertilizers iii) Erosion by water iv) Wind erosion	Use less toxic pesticides Use biological and botanical pesticides Integrated pest management Integrated weed management Crop rotations Monitor the quality of the waters and soils Fractionate the application of soluble fertilizers Place fertilizers Sow leguminous crops Increase the use of organic manures Increase the use of compost Monitor water quality See G Monitor water quality See H Monitor air quality

Possible solutions

i. Lack of soil moisture

If a lack of soil moisture is the main limiting factor, possible solutions are to leave crop residues or to apply mulch to the surface to reduce the evaporation. Windbreaks will reduce the wind velocities and the evaporation rates in zones subject to high winds. A practical guide concerning the establishment of windbreaks may be found in Barber and Johnson (1992). Another possibility is to sow a cover crop during the preceding season and to mow it at least several weeks before sowing the main crop. This provides a mulch cover reducing moisture losses through evaporation. Leaving a straw cover or a killed cover crop on the surface at the time of sowing implies the adoption of systems of conservation tillage, preferably zero tillage, which requires the acquisition of machinery or equipment for direct drilling.

Another solution in well-structured soils is to sow in the dry soil before the rains commence, placing the seed at a greater depth so that it only germinates when there have been sufficient rains to raise soil moisture sufficiently in the seed zone to allow good germination. The seed thus does not germinate from light rains.

ii. Excessive soil moisture

If the problem is excess soil moisture due to the presence of impermeable layers, drainage ditches and canals should be installed. Deep tillage with a subsoiler working transversely to the line of the drainage canals will ease the drainage, and forming raised beds and ridges and planting on the ridge will serve to raise the rooting zone above the soil horizon that is saturated.

If the excessive soil moisture is due to the entry of runoff water from adjacent higher areas, it will only be necessary to install a cut-off drain to prevent the runoff from entering the field.

When the cause is a small irregularity in the topography leading to drainage problems in the depressions, a common problem in soils with deficient drainage, levelling should be undertaken using a light-weight disc harrow (with discs no larger than 22 inches in diameter), with a long-fingered rake hitched behind. It is often advisable to loosen the soil beforehand with a chisel plough. For soils with a heavy texture, a levelling blade may be hitched behind the disc harrow but this method is not recommended for light or medium textured soils, nor is it recommended to use a heavy wooden plank as these tend to pulverize the soil.

iii. Excessively high soil temperatures

If the poor germination is attributed to excessively high soil temperatures, a straw cover or a killed cover crop can be left on the surface to lower the temperatures. It will also be necessary to practise a system of conservation tillage.

iv. Excessively low soil temperatures

When the poor germination is due to excessively cold soil temperatures, the introduction of a system of ridges is a valid option, sowing on the ridges that are devoid of a cover of residues. Another alternative, which does not constitute a conservation tillage system, is to till the soil leaving no surface residues. Both these practices avoid leaving crop residues over the seeding zone, which would lower the soil temperature. The installation of windbreaks is also advisable so as to reduce the wind-chill effect of strong winds.

v. Large and hard clay aggregates

The situation becomes more difficult when poor germination is due to large and hard clay aggregates, which give little contact between the soil particles and the seed. These aggregates normally break down only very slowly under the action of the rain. As a general guide, in conventional tillage systems where aggregates are exposed to the rain, the land preparation should leave clods that are about 5 cm in diameter in clay soils. Later, under the action of the rain, the size of the aggregates will diminish until they reach an optimum size for germination, which varies from 0.5 to 8 mm in diameter.

In most clay soils the consistency changes very rapidly at the start of the rains, from hard in the dry state to friable in a moist state, which is the optimum consistency for tillage, to plastic and sticky when it is wet. The soil will only be in an optimum state of consistency for a very short time, sometimes only for half a day.

In the short period when the soil is in a friable state, harrowing or the use of a rotary cultivator can reduce the size of the aggregates, but the harrow leaves the soil bare, and the rotary cultivator tends to pulverize the soils. It is preferable to use one or two passes of a chisel plough followed by one or two passes with a vibro-cultivator when the soil is at its optimum moisture content. Two lightweight rolls, of the open cage type with angled bars should be hitched behind the vibro-cultivator. Adjusting the pressure on the rolls, can control the degree of disintegration of the clods. Alternatively, one may use heavy clod-breaking rollers with helicoidal bars, which leave more straw cover on the soil surface. It is almost impossible to break down the size of the clods when the soil is dry, and harrowing in this state causes excessive soil pulverization.

Over the long term, the structure of these soils can be improved by leaving the land fallow under pasture for several years. The dense network of the grass roots causes the formation of smaller soil aggregates. However, the beneficial effects only last a short time after cultivating the soil again with conventional tillage systems, perhaps as little as a year. If there are no problems of deficient drainage in these soils, it would be better to introduce zero tillage after the fallow period so as to prolong the beneficial effects.

B. **Adverse conditions for emergence**

Causes of poor plant emergence can be attributed to the formation of surface crusts, particularly in soils with a high content of fine sand, or to the formation of a massive, compacted and hard structure in hardsetting soils when they dry out after a heavy rainfall. The texture of hardsetting soils varies from light to medium.

Identification of soils susceptible to crusting

Soils that have been recently ploughed, cultivated or subjected to traffic, or that are under a cover of crops, weeds or vegetation, will not have surface crusts even if they are susceptible to crust formation.

Sandy soils susceptible to crusting often have grains of clean quartz on the surface. The impact of the rain droplets separates the particles from the aggregates, including the grains of quartz, which generally are very clean due to the low content of clay and humus cementing the grains.

Another field test consists of a simplification of the Emerson test (Cochrane and Barber, 1993). Using a spatula, three to five air-dry aggregates of between 2.5 and 5 mm in diameter are very carefully placed into a receptacle such as a Petri dish, preferably containing distilled water. The stability of the aggregates is then observed to see if they break up, or if the clay fraction disperses partially or completely. Soils that are very susceptible to crusting, break up rapidly during the test and there is dispersion of the clay as a small cloud around the aggregate. Normally, aggregates that do not break up and which only swell, are not susceptible to the formation of crusts.

Identification of hardsetting soils

Hardsetting soils can be recognized by their massive structure and hard consistency when dry, and by the difficulty or even impossibility to cultivate them until their profiles have again been moistened. These soils have been described as "soils that are distubed or indented by the pressure of a forefinger when applied to the air-dry profile face at 0.1 m below the surface of a dry soil" (Mullins *et al.*, 1990).

Possible solutions

i. For surface crusts

Crust formation may be avoided by leaving crop residues on the surface, sowing a cover crop, or by applying a mulch or organic manure over the soil surface. These practices protect the soil aggregates from the energy of the rain droplets, the aggregates do not break up and crusts do not form. In addition, the vegetative cover reduces evaporation and so maintains a higher moisture content in the topsoil, which reduces soil strength. However, the maintenance of a vegetative cover over the surface implies the use of conservation tillage techniques and availability of the respective machinery.

Another solution is making narrow ridges and planting on top of these. Although the crusts still form, their resistance is much less on the crest of the ridge and tension cracks are frequently formed which eases the emergence of the seedlings.

In order to decrease the force that the seedlings need to break through the crust, seeds can be sown at a higher density or a shallower depth. These practices will increase the probability that more plants emerge.

ii. For hardsetting layers

A temporary solution to the problem of hardsetting soils is through tillage to loosen the soil before sowing. In these soils it is practically impossible to undertake tillage when the soil is dry, even with a disc plough as the discs will not penetrate. The soil can be loosened when it is friable using a pass of a disc plough followed by one or two harrowings, but this type of tillage will leave the soil bare and very susceptible to erosion and renewed crust formation. Loosening the massive hardened layers will not prevent the soils from again becoming compacted when they dry out following a heavy rainfall, so restricting the seedlings.

C. **Adverse conditions for root development**

Conditions that are prejudicial for the growth and functioning of roots include soil compaction due to tillage or natural processes, hardsetting soils, large hard aggregates that cannot be penetrated by roots, a deficit or excess of soil moisture, a lack of phosphorus or the presence of toxic elements such as aluminium or manganese.

Identification of soil compaction

The existence of soil compaction problems may be readily recognized by digging a pit and observing the crop roots once these are developed: when the crop is in flower or a later stage of growth. The presence of a hard pan in the lower soil layers is often indicated by the roots growing laterally, by swollen roots occurring above a layer of low porosity, or by their vertical growth being restricted to the first 15 to 20 cm depth. In the absence of a well-developed crop, a simple guide for recognizing compacted layers is to count the number of pores that are visible at a quick glance. If these amount to less than about 20 pores per 100 cm^2 in a vertical profile, then it is probable that the horizon is compacted and the free penetration and development of the roots is restricted.

Determination of the difference between field capacity and plastic limit enables those soils that are susceptible to compaction to be recognized even though they have not been subjected to tillage operations. Those soils with a field capacity that is greater than the plastic limit are more susceptible to compaction, the susceptibility increasing as the difference becomes greater (Barber *et al.*, 1989). Soils with high contents of silt or fine sand and imperfect drainage are often more vulnerable to compaction.

Possible solutions

i. For severe compaction (i.e. recuperation)

   Two crossed passes with a subsoiler when the soil is dry are necessary to loosen compacted layers. The tines should penetrate to a depth of about 1.5 times the depth of the lower limit of the compacted stratum. The reason for tilling so deep is to ensure that the soil between the rows where the tines pass is loosened and the lower limit of the compacted zone is broken up. The tine spacing should be approximately equal to the depth of penetration of the tines so as to ensure that the entire compacted layer is broken up.

ii. For incipient compaction

   Tined tillage is the best way to avoid problems of incipient compaction and the system works well for different soil types, including those with drainage problems and those that are susceptible to compaction. The implements most commonly used are the stubble mulch chisel plough, vibro-cultivator, and the stubble mulch cultivator. More details regarding tined tillage are given in *Chapter 8*.

   Tined tillage combined with controlled traffic is the best soil management system to avoid compaction. In this system, all machinery passes over the same tracks. The soil under the tracks becomes compacted and suitable for traffic, whereas the rest of the land is not passed over by the machinery and so does not become compacted. However, this system requires all the machinery to have the same track width, and requires skill on the part of the tractor drivers.

iii. For hardsetting layers

   The solutions to rooting problems due to hardsetting are identical to those for compaction problems (see Section ii above).

iv. For excessive moisture, lack of phosphorus and presence of toxic substances.

   Solutions for solving problems of excess moisture are dealt with in Section E.ii, lack of phosphorus in Section D.i and the presence of toxic substances in Section D.iv.

D. **Conditions of low fertility and productivity**

Identification of adverse nutritional conditions

As a first step in formulating a soil management strategy it is very important to identify any nutritional problems in the soils. If crop growth is limited by nutritional factors it will not be possible to keep the soils or crops in a good condition. Together with soil analysis, foliar analysis is an important tool that should be used with greater frequency. In order to correctly interpret foliar analysis it is essential to sample the correct part of the crop at the appropriate stage of growth. Often the symptoms of foliar deficiencies serve as useful indicators, but in general they only appear when the deficiencies are already accentuated.

Aluminium is toxic for many crops and it reduces yields when it accounts for 40 percent or more of the effective cationic exchange capacity. There is, however, a great deal of variation in the tolerance to aluminium and manganese between different crops and varieties.

Possible solutions

i. For nutritional deficiencies and/or imbalances

Overcoming nutritional problems requires the application of fertilizers, including organic manure. When applying inorganic fertilizers, it is important to know both the economic application rate and the dose required to attain the maximum yield. The recommended dose, which is a function of these two parameters, will depend on the financial circumstances and the objectives of the farmer. The fertilizer management system should consider the method of application (broadcast or placed) and the number and timing of the applications to maximize efficiency, avoid fixing phosphorus, and avoid excessive applications of nitrogen and other soluble nutrients.

It is also important to apply fertilizers that do not cause acidity so as to avoid chemical degradation of the soils. However, if the soils have a low sulphur content, application of gypsum (calcium sulphate) may be needed when substituting ammonium sulphate, a nitrogenous fertilizer that most acidifies soils but which contains sulphur, with less acidifying fertilizers such as calcium nitrate, calcium-magnesium nitrate, calcium-ammonium nitrate or urea.

In a situation where nitrogen is the limiting nutrient, sowingand incorporating green manure and using legumes in the crop rotation pattern can assist in largely overcoming the nitrogen deficiency. Furthermore, the presence of legumes in the crop rotation will ease the control of graminaceous weeds. It is advisable to use legumes that readily form nodules without the need for inoculation, because of the difficulty of obtaining inoculants and maintaining their efficiency until they are applied in the field. The efficiency of legumes to fix nitrogen depends on an adequate availability of phosphorus in the soil, and one cannot expect a good production from legumes if the soils are phosphorus deficient. For some legumes such as soybean, continued harvesting of the grain will result in a reduction of the overall nitrogen content of the soil due to the harvested grain containing more nitrogen than the quantity that has been fixed.

ii. For poor soils with low organic matter and clay contents

Valid options available for improving the chemical fertility of these soils include applications of organic manures and composts, and agronomic practices such as leaving residues, cover crops, crop rotations that include legumes, deep rooting crops, inter-cropped legumes, the inoculation of legumes and the incorporation of green manure. However, these practices become particularly important for the management of sandy soils with low natural fertility because when the cover crops are well adapted to the zone, they produce large quantities of biomass, which enriches the soil by supplying nitrogen and organic matter. In addition, the increase in the soil's organic matter content increases the capacity both for moisture retention and for retaining nutrients. It is often necessary to apply inorganic fertilizer and lime, in addition to these agronomic practices.

The application of organic manures and compost can be more important for small farmers because of the low costs involved. In contrast, because there are often limited supplies of organic manure and compost and because of the large labour input required, the practice may be more difficult for farmers producing on a large scale. Organic manures and compost are advantageous in containing a wide range of nutrients, even if in small amounts, and they benefit the physical properties of the soil. It is recommended that farmers use organic manure and compost whenever possible.

Although there are nutritional benefits from leaving residues on the soil surface, their high carbon-nitrogen ratio sometimes requires the application of additional nitrogen so as to avoid immobilizing the soil nitrogen by the microorganisms. It will also be important to practise a system of conservation tillage, preferably zero tillage, so as not to incorporate the residues but to leave them on the soil surface.

iii. For soil with serious leaching problems

In zones where there are serious leaching problems, a cover crop with deep roots will recycle nutrients from deeper layers which cannot be reached by the roots of the majority of crops. Suitable cover crops are *Crotalaria spp., Glycine wightii, Centrosema macrocarpum, Cajanus cajan, Panicum maximum* var. Tobiata and *Pueraria phaseoloides.* The cover crop can be grown during the preceding season, after harvest, or within the same season as the main crop. However, if the two crops are grown at the same time, there may be problems of competition between the cover crop and the commercial crop, depending on the sowing dates and growth rates of the two crops.

One may also take advantage of deep-rooting crops such as perennial crops in pure systems, or crops such as cassava and banana in association with shallower rooting crops, which will reduce the loss of nutrients through leaching. Another similar but more systematic practice is that of alley cropping where the bush and tree species which form the alleys have deep roots. Often, the system of alley cropping is referred to as an example of a sustainable system, nevertheless, despite the technical advantages of the system, there has generally been little adoption by farmers mainly due to the overall cost and opportunity cost of the manual labour needed for pruning the trees. It would appear that alley cropping systems would be more acceptable to non-mechanized farmers in hillside areas where there are erosion problems and high pressure on the land, but where plot sizes are not very small.

iv. For soils with toxic levels of aluminium or manganese

The best option is often to change the crop variety for one that is more tolerant to aluminium. In Brazil, there are many varieties of maize, rice and soybean that are tolerant to high concentrations of aluminium. Alternatively, one could change the crop for one that is more tolerant to aluminium.

Where there are no varieties tolerant to aluminium, lime or dolomitic limestone can be incorporated into the soil to neutralize and replace aluminium with calcium and magnesium.

This practice is applicable where the cost to buy and transport lime is low, and where the toxic concentrations of aluminium are in the surface layers.

Application of organic manure can also have a beneficial effect due to the formation of aluminium-organic complexes, which reduces the activity of aluminium in the soil solution. Enhanced applications of phosphorus can also reduce the toxic effects of aluminium.

When high concentrations of aluminium are found in the subsoil, it is more difficult to neutralize it due to the low solubility of lime and its slow movement into the deeper layers. In these cases, gypsum can be applied, or even better, gypsum mixed with lime because the gypsum is soluble and the calcium in the gypsum more rapidly replaces the aluminium in the lower layers.

v. For soils infested with weeds, pests or diseases and "tired" soils

Crop rotation overcomes or reduces problems due to weeds, diseases, insects, loss of fertility and the structural degradation of the soil. For this reason, it constitutes an essential element for sustainable agricultural systems. In order to achieve these benefits, the crop rotation must take account of the following elements:

- sowing broad-leaved crops (for example, soybean, sunflower and beans) before and after graminaceous crops (such as maize and sorghum) so as to allow good weed control;
- sowing legume type crops before other crops so that the latter crops can benefit from the fixed nitrogen;
- presence of crops that supply large quantities of residues, which do not readily decompose (for example, maize, grain sorghum, sunflower or cotton) so as to maintain or increase the organic matter content of the soil (Table 39);

TABLE 39
Straw production by different crops, and their classification in terms of an index of the degree of organic matter supply to the soil (Barber, 1994)

Crop (season)	Straw Kg / ha	C/N Ratio	Grain/straw Ratio (by weight)	Index of the degree of organic matter supply to the soil
Soybean (summer)	1 570	22	1.56	2
Maize (winter)	3 760	40	0.51	7
Sorghum for grain (summer)	3 600	32	0.82	7
Cotton (summer)	3 520	22	0.29	7
Soybean (winter)	900	22	1.56	2
Wheat (winter)	970	75	1.70	3
Sorghum for grain (winter)	2 680	32	0.82	5
Beans (winter)	900	26	0.87	2
Sunflower (winter)	3 590	33	0.34	7
Fallow crops				
Crotalaria juncea (winter)	7 590	19	-	15
Avena strigosa (winter)	3 010	28	-	7

- a sequence of crops that do not act as hosts to the same diseases and/or pests.

The most appropriate crop rotation for a certain specified zone will depend on a number of factors such as the crops for which there are markets, soil types, management systems, climate, weeds, diseases and pests. It will be necessary to identify for each zone, the rotations that are technically, economically and socially the most acceptable. The characteristics of two-year rotations that have been identified as recommended for well-drained soils and for the crops and conditions of the farmers in the sub-humid zone of Santa Cruz, Bolivia are shown in Figure 59. Some characteristics of rotations that are not recommended for the same zone, and reasons why, are shown in Table 40.

TABLE 40
List of crop rotations and sequences that are not recommended for the sub-humid zones of Santa Cruz, Bolivia (Barber, 1994)

Conventional tillage	**Tined tillage**	**Zero tillage**
Wheat every winter (*Helminthosporium*)	Wheat every winter (*Helminthosporium*)	Wheat every winter (*Helminthosporium*)
Soybean twice a year (*Anticarsia*, bean bugs, *Diaporthe*)	Soybean twice a year (*Anticarsia*, bean bugs, *Diaporthe*)	Soybean twice a year (*Anticarsia*, bean bugs, *Diaporthe*)
Soybean every year (*Diaporthe*, bean bugs, *Anticarsia*, weevils, N, structure)	Soybean every year (*Diaporthe*, bean bugs, *Anticarsia*, weevils, N)	Soybean every year (*Diaporthe*, bean bugs, *Anticarsia*, weevils, N)
Soybean-Sunflower, Sunflower-Soybean (*Sclerotinia)*	Soybean-Sunflower, Sunflower-Soybean (*Sclerotinia)*	Soybean-Sunflower, Sunflower-Soybean (*Sclerotinia)*
Maize-Sorghum, Sorghum-Maize (*Spodoptera*, graminaceous weeds, N)	Maize-Sorghum, Sorghum-Maize (*Spodoptera*, graminaceous weeds, N)	Maize-Sorghum, Sorghum-Maize (*Spodoptera*, graminaceous weeds, N)
Maize-Wheat, Sorghum-Wheat (*Spodoptera*, N, weeds)	Maize-Wheat, Sorghum-Wheat (*Spodoptera*, N, weeds)	Maize-Wheat, Sorghum-Wheat (*Spodoptera*, N, weeds)
-	-	Sunflower-Cotton (broad-leaved weeds, phytosanitary control of the cotton residues)
-	-	*Crotalaria juncea*-Sunflower (resprouting of *C. juncea)*
Sorghum-Cotton, Cotton-Sorghum (*Spodoptera*)	Sorghum-Cotton, Cotton-Sorghum (*Spodoptera*)	Sorghum-Cotton, Cotton-Sorghum (*Spodoptera*, phytosanitary control of the cotton residues)
Soybean-Beans (bean bugs, *Diaporthe*)	Soybean-Beans (bean bugs, *Diaporthe*)	Soybean-Beans (bean bugs, *Diaporthe*)

There is also a phenomenon known as "tired soils" soils where soil productivity is low despite efforts to overcome all the limitations of the soil. Apparently, the only way to raise productivity is by the introduction of a good crop rotation. Possibly, the "tiredness" of these soils is due to the allelopathic effects of the crop sequence.

E. **Adverse edapho-climatic effects for crop development and field operations** (see Figure 60)

Interactions between the rainfall and soil characteristics can result in problems of moisture shortage which adversely affect crop development, or of excess moisture which generates problems for crop growth and field operations such as spraying and the harvesting. Strong winds can also cause problems for crops, causing high evaporation rates of soil moisture, making spraying difficult or even impossible for the control of weeds and insects, or causing wind erosion in sandy soils.

Lack of moisture

Shortage of soil moisture can be caused by a low infiltration rate, high evaporation, low moisture retention capacity of the soil, low or irregular rainfall.

Figure 59
Two year rotations of annual crops, recommended for Santa Cruz, Bolivia for well-drained soils with medium to moderately heavy texture (Barber, 1994)

1st Year		2nd Year		Restrictions And precautions	Degree of supply of organic matter to the soil
Summer	Winter	Summer	Winter		
With maize and cotton in the summer					
1. Maize	Beans	Cotton	Soybean	ZH; CN; CH; R; P; A	Low
2. Maize	Soybean	Cotton	Beans	ZH; CN; CH; P; R; A	Low
3. Maize	Beans	Cotton	Sunflower	ZS; CN; HA; R; P; A	Moderate
4. Maize	Soybean	Cotton	Sunflower	ZT; CN; HA; CH; R; P; A	Moderate
5. Maize	Sunflower	Cotton	Soybean	ZT; CN; HA; CH; R; P; A; N	Moderate
With soybean and cotton in the summer					
6. Soybean	Sorghum	Cotton	Sunflower	ZS; CN; HA; CH; R; P	Moderate
7. Soybean	Wheat	Cottonn	Sunflower	ZS; CN; HA; CH; R; P	Low
8. Soybean	Sunflower	Cotton	Beans	ZS; CN; HA; CH; RH; R; P	Low
With soybean and maize in the summer					
9. Soya	Sunflower	Maize	Beans	ZS; RH; CH; P; A	Low
10. Soya	Wheat	Maize	Sunflower	ZS; LCV; GG; GU; CH; N; P; A	Low
With soybean and sorghum in the summer					
11. Soybean	Sunflower	Sorghum	Beans	ZS; RH; CH; P	Low
12. Soybean	Wheat	Sorghum	Sunflower	ZS; LCV; GG; GU; CH; N; P	Low
With maize and sorghum in the summer					
13 Maize.	Sunflower	Sorghum	Soybean	ZT; CH; P; BG; A	Moderate

CN Suitable for conventional or tined tillage, but *conventional tillage* (disc plough) must be used after the cotton harvest.
LCV Only suitable for *conventional* or *tined tillage*. Not suitable for zero tillage.
ZH Only suitable for *humid zones* where winter soybean may be grown.
ZT Only suitable for *transition zones* where winter sunflower and winter soybean may be grown.
ZS Only suitable for *dry zones* where winter sunflower may be grown.
G Danger of graminaceous weeds in wheat. It requires the application and incorporation of pre-seeding graminicides and selective post-emergence graminicides for conventional tillage and of selective post-emergence graminicides for zero tillage.
GG Danger of graminceous weeds in graminaceous crops. Sunflower, soybean and beans after graminaceous crops require the application of pre-seeding and post-emergence graminicides for conventional and tined tillage and of post-emergence graminicides for zero tillage.
GU Danger of graminaceous weeds in the second graminaceous crop. It requires mechanical control through *cultivations*
HA Danger of *broad-leafed* weeds in the sunflower following cotton, or of resprouting of the sunflower in the cotton. It requires cultivations and manual hoeing.
A Danger of a residual effect of *Atrazine* in the following crop. Do not apply Atrazine in maize.
RH Danger of root rot due to *Rhizoctonia solani* in the beans-soybean and the soybean-beans rotation sequences. It requires the application of fungicides.
MH Danger of *Mustia hilachosa* caused by *Thenatephorus cucumeris* and root rot caused by *Rhizoctonia solani* when beans are sown for two winters in succession; requires the application of fungicides.
R Danger of *ramulosis (Colletotrichum gossypus)* in susceptible varieties such as Guazuncho, which are susceptible to this disease. Other more resistant varieties should be sown such as stoneville.
C Danger of *canker (Diaporthe phaseolorum)* in susceptible varieties such as Cristalina. Sow other more resistant varieties such as Doko.
N Danger of nitrogen deficiency. It could be advisable to apply nitrogenous fertilizer to maize, sorghum and sunflower that follow a non-leguminous crop.
CH Danger of infestations with bugs, particularly *Piezodorus guildinii*. It requires integrated pest management techniques to control *Anticarsia* through the application of residual and systemic insecticides and biological selective insecticides that do not kill the natural enemies of the bugs.
P Danger of attack by *pests* such as *Conotrachelos denieri* aphids and thrips in cotton, and spodoptera and aphids in graminaceous crops. It requires treatment of the seeds.
S Danger that the *seeds* from the fallow crop regrow during the following crop. It is imperative to eliminate the fallow crop when the grain is still at the milky stage.

Possible solutions

i. For soils with moisture deficit due to low infiltration or high evaporation (where there are residues)

For situations where low infiltration or high evaporation limit the availability of soil moisture, the best solution is to leave a surface cover of crop residues. The residue cover will increase the infiltration rate and reduce the evaporation of moisture from the soil. The options available are to leave the residues from the preceding crop, to apply a mulch or

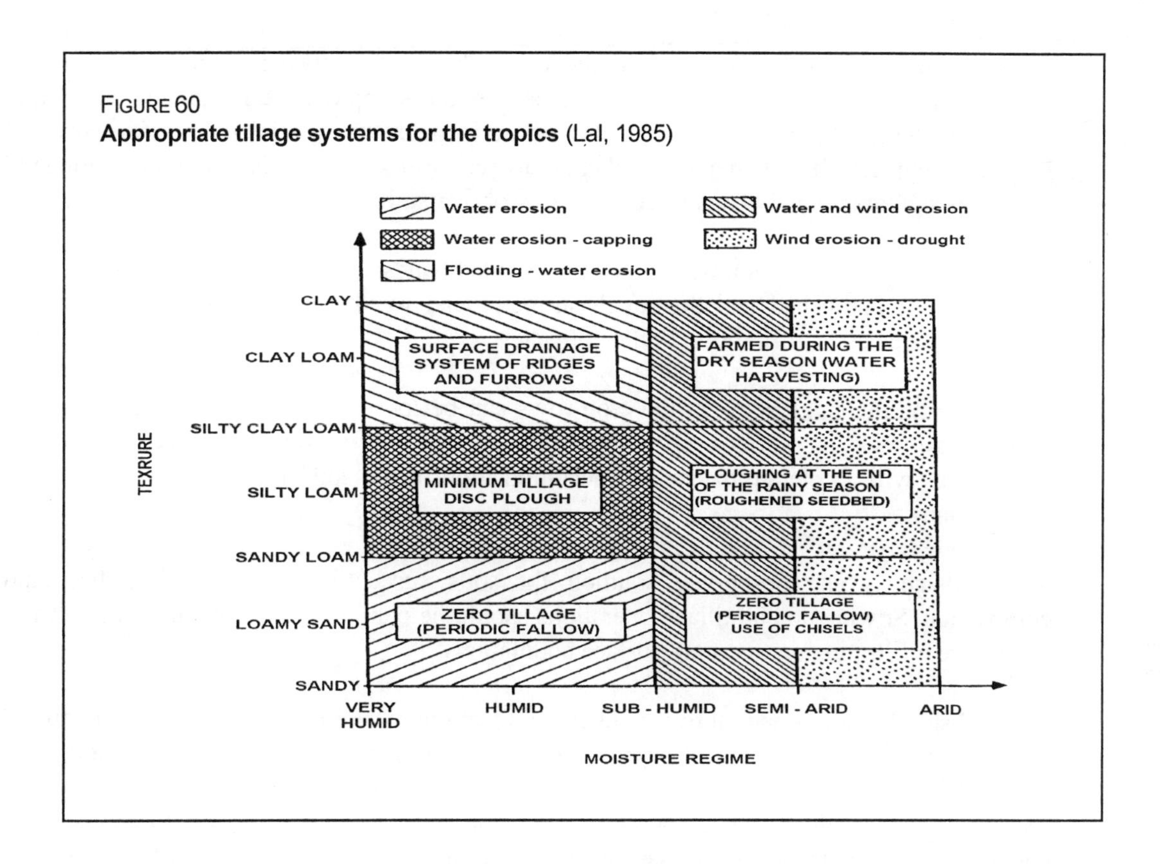

FIGURE 60
Appropriate tillage systems for the tropics (Lal, 1985)

organic manure, or to sow a cover crop. The presence of a vegetative cover on the surface implies the use of conservation tillage techniques, preferably zero tillage, which will leave more residues on the surface and so reduce evaporation more than for conservation tillage systems. In addition, the control of weeds using herbicides or by tillage that does not incorporate the weeds, such as with a stubble mulch cultivator, will maintain a better residue cover. Another option is to change the crop variety for one that is more resistant, or can better escape the periods of drought.

The sowing dates and the date when the cover crop is to be eliminated should be fixed so as not to cause an excessive reduction of soil moisture, which would affect moisture availability for the following crop.

ii. For soils with moisture deficit due to evaporation aggravated by high wind velocities.

Windbreaks are necessary in areas characterized by strong winds so as to reduce the wind velocity and thus reduce evaporative losses to the air.

iii. For soils with moisture deficit due to low infiltration and high evaporation (where there are no residues)

It will be necessary to use physical solutions in areas where there are moisture losses due to poor infiltration and high evaporation. The lack of residues could be attributable to low production levels of straw, or to its use for other purposes such as forage, or because it is eaten by termites.

Constructing narrow tied ridges favours the accumulation of rainwater in the furrows and its subsequent infiltration at the same place. This method is appropriate for manual or animal traction systems but tied ridges can also be constructed by specialized tractor powered ridging machinery. It is a method applicable to semiarid zones but is not suitable for land with slopes greater than 7 percent due to the risk of overspills.

Strip tillage is another option where the narrow tilled bands are sown. The areas between the strips are left untouched and the rough surface, dead weeds and crop residues from the preceding crops assist infiltration of the rainwater.

Alternatively, the areas between the crop rows can be loosened with cultivators after each rainfall to break up surface crusts and increase the infiltration. However, numerous cultivations will accelerate the rate of decomposition of the soil organic matter (biological degradation).

Tined tillage is an appropriate solution for soils susceptible to crust formation and compaction. Sowing and all tillage operations should be parallel to the contour lines so as to encourage infiltration in the small surface undulations.

The soil should be loosened through tillage in situations where the low infiltration is due to the presence of massive and hardsetting soils. In West Africa there is evidence that ploughing at the end of the rainy season loosens the soil profile and favours infiltration over the next season. However, this method causes soil inversion that could give rise to physical or chemical problems if the subsoil has undesirable characteristics. For these soils, it would be better to use tined tillage so as not to invert the soil.

iv. For soils with moisture deficits due to a low moisture retention capacity

The moisture retention capacity of sandy soils can be increased through practices that increase their organic matter content, such as incorporating organic manure or green manures, or by planting a cover crop.

In soils where impermeable layers impede water percolation to deeper layers, subsoiling that breaks up these layers will increase the amount of moisture retained. In clay soils, subsoiling can also increase moisture retention capacity by increasing the surface area available for absorption of the moisture.

Another possible option is to change the crop variety for one that is better adapted to, or that can escape, the periods of drought.

In zones where agriculture is not too intensive, another practice to increase the availability of soil moisture is by introducing periodic fallows. In areas where it is not feasible to sow two crops a year, leaving the land to rest during one of the seasons will allow the accumulation of moisture for the following crop. However, it is necessary to control any growth in the fallow field so as not to exhaust the accumulated moisture supply. Periodic fallows of several years will also improve infiltration and moisture retention capacity by increasing the organic matter content and infiltration rate of the soil. When conventional tillage is practised after a period of fallow, the beneficial effects will generally last for less than a year. It may be possible to obtain better results with conservation tillage.

Sprinkler, gravity or drip irrigation obviously contribute towards overcoming periods of drought but can involve considerable investment costs, depending on the system and its scale.

v. For soils with moisture deficit due to runoff and low moisture retention capacity

Another way to increase infiltration is by changing the slope of the land. The construction of terraces will increase the infiltration of rainfall but the high costs involved mean that the method is only appropriate for high value crops. The construction of individual terraces will also increase infiltration and moisture retention capacity and is suitable for orchards.

Excessive moisture

When soils have a moisture content that is higher than the optimum, field operations with machinery and implements will increase the risks of soil compaction. This situation can occur during primary tillage, spraying or cultivation operations, or during harvest. The excessive moisture may be due to an ingress of runoff water, the presence of impermeable layers, or a high water table.

Identification of problems due to excessive moisture

The presence of colours or at least 10 percent of grey-coloured mottles in the soil indicates that anaerobic conditions due to excessive soil moisture have persisted for a number of months.

Possible solutions

i. For excess moisture due to runoff

The only solution is to construct cut-off ditches. It is very important to ensure that the discharge of the diverted water does not cause problems at the point of discharge.

ii. For excess moisture due to a high water table or impermeable layers.

If the problem is due to an impermeable layer within the first 60 cm of soil depth, a system of shallow open ditches should be installed. However, if the impermeable layer is at a depth between 80 and 100 cm, deeper drainage ditches will need to be installed with their bases located above the impermeable layer. In addition, clay soils will need a narrower spacing between the ditches. The practice of deep tillage with a subsoiler working in a direction perpendicular to the line of the ditches will assist drainage.

The construction of broad cambered beds up to 20 to 30 metres wide and slightly sloping, will ease drainage of the excess moisture towards the ditches and raise the rooting zone above the soil horizon that is saturated. The broad cambered beds may be constructed with a conventional plough, working towards the middle of the bed or alternatively, a levelling blade may be used.

Forming narrow ridges and planting on the ridge also serves to overcome problems of a high water table.

When the excess moisture is due to compacted layers, deep tillage with a subsoiler to break up the hard pan will improve the drainage.

Strong winds

Strong winds can cause not only problems of wind erosion and crop damage, but can also occur at critical times when spraying must be carried out. A delay of only a few days in the application of pre-emergent herbicides to maize due to strong winds can increase the risks of infestation with graminaceous weeds.

Possible solutions

As has already been described, the installation of windbreaks can limit the harmful effects of strong winds.

F. **Lack of biological activity**

Possible solutions

The lack of biological activity may be due to a lack of crop residues, "tired" soils or to the application of toxic pesticides.

i. For soils with little crop residue cover

In order to increase biological activity in the soil, a persistent cover of dead vegetative material is needed, which can be obtained by leaving crop residues on the surface, applying mulch or organic manure, and by sowing a cover crop. The active presence of earthworms in a soil needs a supply of dead vegetative material throughout the season when the soil is moist. If there is good soil moisture but a lack of dead vegetative material, the earthworms will move elsewhere.

The only way to maintain a residue cover is through conservation tillage, particularly zero tillage. Conventional tillage involving soil inversion will not leave a sufficient quantity of residues on the surface.

Another option is to increase crop residue production by increasing the chemical fertility of the soil through the application of organic manures or fertilizer.

Alternatively, crops and varieties that produce large amounts of biomass may be sown, and which preferably produce residues that do not decompose rapidly. The resistance of residues to decomposition varies according to the carbon/nitrogen ratio, the amount of lignin, polysaccharides and the lignin/nitrogen ratio. Table 39 shows a number of annual crops classified according to an index of the quantity of organic matter that they supply to the soil.

ii. For "tired" soils

It is probable that the phenomenon of "tired" soils can be overcome through changing the crop rotation. This has been discussed above in Section D.

iii. For soils receiving high concentrations of pesticides

The application of massive amounts of non-specific pesticides reduces the biological activity of soils. This activity can be better maintained through the application of biological and botanical pesticides, integrated pest management, and by the application, when necessary, of selective pesticides. Integrated control of weeds can also reduce the quantities of herbicides that are needed.

G. **Erosion by water**

Identification of the presence of water erosion

Careful observation of the microtopography of the soil surface will reveal any evidence of the processes of water erosion. These may include, for example, flat and smooth areas caused by sheet erosion, small rills and channels due to rill erosion, etc. Root exposure, the formation of small pedestals, and the accumulation of sediment behind barriers are also indications of active water erosion processes.

Possible solutions

Problems of water erosion can be considered as the result of low infiltration due to a lack of soil cover or surface roughness on sloping land. The presence of cover which is in contact with the soil and surface roughness will give more time for rainwater to infiltrate and so reduce the risks of erosion. Tillage practices that increase surface roughness are only suitable for gentle slopes. Some practices, such as levelling or drastically reducing the land slope, will hinder the initiation of runoff, whereas other practices only trap the runoff after it has had an opportunity to cause erosion.

i. To increase infiltration through increasing soil cover.

Increasing the soil cover involves increasing residue production by means of all practices of fertilization and manuring that increase soil fertility, through the introduction of crops or varieties that produce greater quantities of residues, or more resistant residues, and through sowing higher plant population densities.

Sowing cover crops, intercrops or fallow crops will also increase soil cover and provide more crop residue to protect the soil. Another option is to apply materials such as mulch and organic manure.

In order to reduce the losses of crop residues, the straw and stubble should not be burned and grazing should be reduced to a minimum in livestock zones. In order to reduce grazing, the fields should be fenced and alternative sources of forage produced.

It is important to practice systems of conservation tillage. Zero tillage does not disturb the soil and so all the residues remain on the surface. Tined tillage leaves a smaller amount of crop residue on the surface but this is often sufficient to restrict water erosion although it is often insufficient to achieve an optimum level of biological activity. In mechanized systems, the use of cultivators for weed control is better than a disc harrow because the cultivators uproot the weeds and leave them on the surface. In contrast, disc harrows partially invert the soil, burying the weeds and leave the soils less protected. A better practice is to control the weeds with herbicides.

For manual tillage systems, the presence of stones on the soil surface will act as a cover and protect the soil. It is better to leave them in place rather than remove them to construct stone barriers.

ii. To increase infiltration through increasing surface roughness

Every practice causing irregularities parallel to contour lines will assist infiltration of the rainwater. Practising tillage and sowing parallel to the contour lines will increase infiltration but these practices will only function on gentle slopes.

Systems of beds and furrows, such as ridge tillage, reduce the erosion. However for slopes steeper than about seven percent there is an increased risk of collapse and overspill, which can cause very serious erosion due to the downhill flow of the water that has accumulated in the furrows.

Strip tillage leaves undisturbed all the area between the crop rows. If there is a protective cover of straw and dead weeds, this will have the effect of slightly increasing surface roughness and improving infiltration. A variation of this practice involves ploughing after the rainy season, which will increase surface roughness; secondary tillage is then undertaken only along the strips where the crop is to be sown. In this manner, the zone between the rows is rougher but the soil is left with little cover. In areas where crop residues are not available, this practice may be accepted although it does not protect the soil and does not favour biological activity.

iii. To reduce the quantity and velocity of runoff

In manual or animal traction systems, alley cropping parallel to the contours will reduce the velocity of the runoff and through this soil losses, providing there is sufficient vegetative cover, both dead and alive, to form a dense barrier in contact with the soil. Only those materials in contact with the soil can reduce runoff velocity and cause sediment to be deposited. In order to accelerate the development of the barriers, all the clippings from the pruning operations should be placed along the upslope side of the row of trees and parallel to the contour.

TABLE 41
Species that are used or have been found to be promising as live barriers

Brachiaria brizantha *Brachiaria decumbens* *Andropogon gayanus* *Andropogon citratus* *Phalaris sp* *Pennisetum purpureum* (Elephant grass) *Saccharum oficinarum* (Sugar cane) *Panicum coloratum* (Makarikari grass)	*Vigna unguiculata* (Cowpea) *Cajanus cajan* (Pigeon pea) *Ananas comosus* (Pineapple) *Crotalaria sp.* *Leucaena leucocephala* *Agave letonae* (Salvador henequen) *Gliricidia sepium* *Sempervivum sp.* (stonefruit)

Live barriers planted along contour lines will also reduce runoff velocity provided they form a dense barrier. Perennial or semi-perennial plants may be used, planted parallel to the contour and with a spacing between the barriers that depends on the slope. The species should be adapted to the zone, should provide additional benefits to the farmer such as grass, forage, fruit, spices or grain, and they must not invade, shade or compete with neighbouring crops. Table 41 presents a list of species that are used or have been found promising, as live barriers in El Salvador. Table 42 presents recommendations for the spacing between the barriers according to the slope and crop type, together with other recommended practices with which they should be combined for non-mechanized hillside farmers in El Salvador.

TABLE 42
Guide to the selection of soil conservation practices for different crops and slopes in El Salvador

Crop	Slope %	Conservation practices **
Horticulture	0-5	Mulch, sowing along contours, strip cropping
	5-10	Mulch cover, sowing along contours, strip cropping, contour beds, stone walls, bench terraces
	>10	Horticultural crops are not recommended, but if this is unavoidable, apply the practices listed for slopes of 5-10%
Basic grains	0-10	Do not burn but leave the residues, direct drill, sow along the contours, control grazing, establish live fences along contours, inter-crop with legumes.
	10-20	Do not burn but leave the residues, direct drill, sow along the contours, control grazing, establish live fences along contours, inter-crop with legumes, establish live barriers approx. every 12 m, establish hillside ditches*.
	20-35	Do not burn but leave the residues, direct drill, sow along the contours, control grazing, establish live fences along contours, inter-crop with legumes, establish live barriers approx. every 8 m, establish hillside ditches*.
	35-50	Do not burn but leave the residues, direct drill, sow along the contours, control grazing, establish live fences along contours, inter-crop with legumes, establish live barriers approx. every 6 m, establish hillside ditches*.
	>50	Basic grains are not recommended but if this is unavoidable, apply the practices listed for slopes of 35-50%.
Fruit trees	0-10	Mulch, leguminous cover crops, live fences along contour lines.
	10-20	Mulch, leguminous cover crops, live fences along contour lines, live barriers spaced at approx. 12 m, hillside ditches*, individual terraces.
	20-35	Mulch, leguminous cover crops, live fences along contour lines, live barriers spaced at approx. 8 m, hillside ditches*, individual terraces.
	35-60	Mulch, leguminous cover crops, live fences along contour lines, live barriers spaced at approx. 6 m, hillside ditches*, individual terraces.
	>60	Fruit trees are not recommended but if this is inevitable, apply the practices listed for slopes of 35-60%.

* Only recommended for soils with a degraded surface structure.
** Land subject to runoff from road, culverts or from steep slopes requires cut-off drains.

It is more difficult to obtain and maintain a residue cover over soil in semiarid climates due to the lower biomass production in the lower rainfall regime, which is often accompanied by an intense termite activity. There are also situations where, due to economic factors, the farmers can only sow crops that produce small amounts of straw such as soybean or beans, without the possibility of sowing cover crops. The combination of these factors, associated with high temperatures, makes it difficult to produce and maintain an adequate cover of residues over the soil. Under these circumstances, the reduction of runoff velocity requires a combination of live barriers with physical structures such as hillside ditches and stone walls.

Stone walls are not recommended for manual systems because they remove the surface stones needed for their construction, so removing the cover that was encouraging infiltration.

Blind ditches are pits that are normally constructed in perennial crops, which trap the runoff according to their location, size and the spacing between them.

When the erosion problems are due to runoff entering from areas outside the field, diversion drains should be constructed to capture the runoff and to carry it away from the field to a safe disposal point.

All practices that trap runoff will reduce overall problems of erosion, but will not prevent erosion occurring between one structure and another.

The construction of bench terraces, orchard terraces and individual terraces with level slopes will have the effect of impeding the initiation of runoff.

H. **Wind erosion**

Identification of the presence of wind erosion

Occasionally one may observe the accumulation of sand deposits behind barriers or hedges, which have been carried there by the wind, or trees leaning in the direction of the prevailing wind (although this does not necessarily mean that there is wind erosion of the soil). However, there are generally few signs of wind erosion outside the season for strong winds.

Possible solutions

i. For soils with a lack of cover

In the same way as controlling water erosion, the presence of a residue cover, weeds, stones, mulch or manure can all help to protect the soils from wind erosion. It follows that systems of conservation tillage that leave residues anchored or fixed on the soil surface will reduce the risks of wind erosion. The application of herbicides in zero tillage systems, or the use of tined cultivators in tined tillage systems, will leave the majority of weeds over the soil as a protection.

ii. To reduce the wind velocity

The most common practice to control wind erosion is the construction of windbreaks established perpendicular to the direction of the prevailing winds. The spacing between the windbreaks should not exceed ten times the expected height of the trees in the windbreak. In order to avoid turbulence, the windbreaks should have a level of permeability of about 40 percent.

Ridging and sowing in the furrows across the direction of the prevailing winds will also reduce wind velocity due to increasing surface roughness which reduces the effects of erosion.

I. **High production costs**

The main causes of high production costs are attributable to the costs of labour, machinery, pesticides and fertilizers. Possible solutions are summarized in Table 38 and all of these have already been dealt with in the preceding sections.

J. **Environmental pollution**

The main causes of pollution of the environment are linked to the use of toxic pesticides, the loss of soluble fertilizers, erosion by water and wind erosion. Possible solutions are summarized in Table 38 and again, these have all been dealt with previously.

REFLECTIONS CONCERNING THE SELECTION OF SOIL MANAGEMENT TECHNOLOGIES

This section has dealt with a wide variety of soil management systems and it is opportune to emphasize some points. Firstly, it is always advisable to consider a variety of technological options because rarely can one solve real problems with the introduction of a single practice. Secondly, the elements of solutions presented in this document should be considered as a guide to arrive at a combination of the most appropriate practices for a given situation. What may work well in one place may not be the most appropriate for another location, despite similarities that might exist. However, if one always remembers the guiding principles "increase soil cover, increase the organic matter, improve the infiltration and retention of moisture, reduce runoff, improve rooting conditions, improve fertility and productivity, reduce production costs, protect the field and reduce environmental pollution", then in general terms, overall positive results will be achieved.

Chapter 17

Participatory planning in the execution of soil management programmes

In order to execute an effective and successful participatory programme for improving the soil through conservation practices, it is necessary to develop specific strategies and methodologies and to apply and validate them on the basis of a feedback process originating in the very actors involved in the work.

Various soil conservation programmes and projects were executed in the decade of the 1950's in Santa Catarina. Both positive and negative experiences occurred. For example, a programme was aimed at local farmers, mainly on hilly land, which focused principally on erosion control through mechanical practices such as terracing. The result was only limited adoption of the practices and unfortunately, these were quickly abandoned. There were no financial incentives and there was no lack of training on the part either of the technicians or the farmers involved. The cause for the failure was that the recommended practices for controlling the erosion were directed basically towards the effect and not the cause of the problem as hardly five percent of the erosion by water was caused by surface runoff.

In reality, the problem of hydraulic erosion in the southern part of Brazil is directly related to the lack of soil cover. Towards the end of the 1970s this aspect was eventually duly considered, and this determined fresh objectives for the research and rural extension work. During this time, spasmodic and isolated field activities continued to be undertaken. Starting in 1984 and based on experience gained in the State of Paraná, a soil conservation project was executed in Santa Catarina, which considered hydrographic micro-catchment areas as the units to be used for planning purposes. The work commenced with three micro-catchment areas, was expanded to 17 in 1985 and later to 68 as a National Programme was created. Later there was a period of recession in project activities, but these gained momentum again in 1990 with the collaboration of the Programme for the Recovery, Conservation and Management of Natural Resources in Hydrographic Micro-Catchment Areas, known simply as the Proyecto Microcuencas/BIRD. The project was assisted by a loan of 33 million dollars from the World Bank and its objective was to reach a total of 520 micro-catchment areas by the end of 1997.

These years of experience accumulated by the Enterprise for Agricultural Research and Rural Extension at Santa Catarina have contributed to the impact of the soil management

V. Hercilio de Freitas
Enterprise for Agricultural Research and Rural Extension of Santa Catarina (EPAGRI)
Santa Catarina, Brazil

programmes and the results were so significant that a second Project was prepared for initiation in 1998.

However, the strategies and methodologies are under constant review through a feedback process, which is achieved through the current participation of more than 300 technicians in project execution. In this way, new perspectives are presented in order to perfect the work. For this reason, the proposals presented below cannot be considered as definitive, rather they are subject to a continuous process of updating.

MICRO-CATCHMENT AREAS AS PLANNING UNITS

The planning for adequate soil management programmes should be considered within an overall programme of rural development. There is recognition on the part of governments, national and international organizations that by limiting development to certain specific components, the more global problems become only partially solved. A rural development plan can only achieve its objectives if one considers the existing land, the land use capability and its ability to produce food, wood and other products and benefits of use to mankind. The production systems are designed on the basis of economic parameters and with the objective of increasing the profits of rural families. The plan necessarily includes maintaining soil productivity over the long term and re-establishing the original balance that has a direct influence on the circulation of natural water (the hydrological cycle). The plan must also take into account the existing infrastructure in terms of the market, transportation, storage facilities and all the elements that are comprised within the system.

Planning activities that aim to reduce the degree of deterioration of the physical, social, economic and environmental aspects will be effective only if these are undertaken within natural limits. That is to say, activities should be within ahydrographic micro-catchment area or in each of its independent tributaries such as valleys or sub-valleys.

For some considerable time, the units for planning agricultural activities have been composed of rural communities and more specifically, the land of individual private farmers (political planning units). The use limits generally do not coincide with those established by the forces of nature because the main effects of these forces affect the entire geographical complex where actions are undertaken. For this reason, the traditional political units are now being abandoned as the sole basis for planning.

The new concept for the planning unit is based on the following principles:

1. degradation of agricultural land generally occurs independently from the political and administrative divisions;
2. once the basic elements of a plan for a hydrographic basin are known, effective work can be organized and undertaken on a smaller scale. Work commences in one micro-catchment area and upon completion of this, attention is then turned to another micro-catchment basin and so on until the entire basin has been covered;
3. management of hydrographic micro-catchment basins implies the rational use of the soil and water, trying to optimize and sustain production with a minimum risk of degrading the environment;

4. the hydrographic micro-catchment basin forms the basis for a physical planning unit, whilst the community continues being, more than before, the nucleus and the basis for the decision making process.

IMPLEMENTATION OF PROGRAMMES AND PROJECTS

Over a long period of time in Santa Catarina, the programmes were practically implemented with minimum economic resources allotted for the maintenance of the equipment and for the production of training material. As from 1990 when the Proyecto Microcuencas/BIRD commenced, the assistance in resources reached a total of 78.6 million US dollars of which 17.4 million were allocated exclusively for rural extension activities planned in the 520 micro-catchment basins, so constituting a large scale programme. The work structure gave place to the creation of a true team spirit between the different hierarchic levels. In reality, small programmes can have some advantages such as being more creative and can allow necessary changes without bureaucratic difficulties and tend to be more sensitive to user needs. But no programme, regardless of its size, can do without a well prepared and motivated technical field staff, which knows the families involved, which treats them with respect and which is well looked upon.

Furthermore, in order for the programme to be successful it needs to be guided through a broad understanding of the needs, motivations, values and points of view of the rural families. The technicians involved in the programme must have great sensitivity in order to be able to deal with the delicate balance between:

- the value of the changes and the respect for the traditional values of the community;
- the need for excellence in the work and the need for the rural families to have freedom to learn from their own errors;
- the need for extraordinary motivation in work and the danger to suffocate enthusiasm through excessive work.

In order to achieve success, the participating technicians above all need to be motivated with a genuine desire to struggle for the wellbeing of others and to stimulate the development of an innate capacity of the rural families as regards their self-determination and their ability to undertake tasks on their own.

The concept of the rural extension worker grows from these principles so that the technician becomes a facilitator in the execution of whatever programme.

OBJECTIVES OF A PROGRAMME AND PROJECT

A programme must include training and motivate the farmers so that they themselves transmit the new practices to other farmers. At the same time, they must be shown that they can discover and test innovations by themselves. In a few words, the objective of the programme should not be the development of the farming practices of the rural families, rather it should be to teach them a process by which they can develop innovative farming practices by themselves.

ENTHUSIASM AS THE DRIVING FORCE FOR DEVELOPMENT

Enthusiasm can be defined as motivation, determination, desire, delivery, compromise, inspiration and love of the work.

When enthusiasm is lacking:

- nobody participates in the meetings;
- co-operation between neighbours is almost impossible and non-existent;
- the extension workers seem to be incapable of convincing the farmers.

When enthusiasm exists:

- the farmers do not measure the efforts required to participate in meetings;
- innovations become spontaneous between one farmer and another.

Factors that can assist in creating enthusiasm

- the families must have a strong desire to overcome the problems that the programme is proposing to resolve;
- the families must believe that they are capable of solving the problems according to their own possibilities;
- the families must be involved in the planning process to the stage that they feel also that the success of the programme is due to them;
- the families must feel that the programme belongs to them;

Enthusiasm increases when the families have:

- freedom to establish their own work objectives;
- freedom to be creative;
- the opportunity to work in a climate of companionship and mutual assistance;
- the opportunity to continue to learn new subject matter that interests them and to find solutions for other needs;
- recognition, gratitude and positive feedback received from other families and from the technicians involved in the programme.

SUCCESS – THE SOURCE OF ENTHUSIASM

None of the statements mentioned previously would inspire greater enthusiasm if an additional and absolutely essential ingredient were to be missing: rapid and recognizable success.

Organizing competitions or handing out prizes to outstanding farmers cannot stimulate interest and enthusiasm. If a technology brings about success, the prize becomes superficial and useless.

PARTICIPATION - THE ROAD TO FOLLOW

Constructive participation

Just as enthusiasm and positive effort can prevent a programme becoming paternalistic, so increasing participation is the road that the programme should follow. This is the opposite to "Doing things for the people" and "Participation of people with nothing else to do". This participation has to be real by definition and throughout the programme execution.

The most important reason for participation is that it ensures the permanence of the progress and of the positive results of the programme. Through this, the families learn to plan, to find solutions to their problems, to teach other people and to organize themselves to work as a group. Through participation, skills are acquired to give and to receive orders within an organization and lessons are learnt as to how to correct companionship without damaging it. These are essential acts so that the people can successfully create and manage their own organizations. Participation can lead to gaining self-confidence, pride and satisfaction from the advances made and to developing creativity so as to continue improving life within the community.

This participation is the essence of self-development or descriptively, "a process through which people learn to become masters of their own lives and to solve their problems".

It follows that development occurs when people are obtaining:

- self-confidence;
- motivation and character;
- sufficient knowledge to solve their problems through their direct involvement in attacking them and solving them.

Because of this, two conclusions may be immediately deduced:

1. presenting and doing things for the people can in no way be called development;

2. the development process by which families learn, create and organize themselves is much more important than the fact that the soil is recovered, that there are no erosion losses or that the maize is greener or their pockets are full of money.

Although both things are important and should occur simultaneously, the "how it is done" is much more important than the "results are as follows". The "how it is done" must necessarily include constructive participation.

DESTRUCTIVE PARTICIPATION

This occurs when in a particular programme, certain people from the community assume control and the remainder have to be submissive instead of giving their opinions and participating in the planning and execution process. Often it is the very lack of experience in making group decisions that causes disagreements, generates the formation of factions and causes organizations to disintegrate. It may also happen that a decision that has been carefully taken, does not lead to success. It is therefore essential that participation should be as constructive as possible.

HOW TO IMPROVE CONSTRUCTIVE PARTICIPATION

1. It is necessary to recognize that constructive participation is learnt gradually. Some organizations, in order to avoid a suffocating paternalistic attitude and the attitude that "the Technician knows everything", soon pass to the other extreme of doing as little as possible. One cannot expect rural families rapidly to learn to resolve their problems. It is necessary to seek a balance, at least at the start of the programme. Group independence does not imply the total absence of the technician.
2. Examples of two recognizable and rapid success are essential near the start of the programme. Success attracts highly motivated leaders for a constructive participation, it creates links of friendship between those involved in the programme and stimulates an exchange of experiences between neighbours and friends.
3. All must make a conscious and constant effort so that the families learn to participate. This can be achieved through holding short courses and through constant discussion of the experiences that the programme develops within the community.

HOW TO INCREASE PARTICIPATION

Work with the people and not for the people

This signifies starting the work jointly with the families, analyzing their particular production system and drawing out their knowledge.

1. undertake a diagnosis of the situation with the full participation of all the families from the community so as to highlight their most urgent needs;
2. evaluate priorities according to local resources and possibilities for this. In order to achieve this, the technical facilitator or promoter should have:
   - humility
   - social sensitivity
   - willingness to learn with the people
   - knowledge of the social context
   - knowledge of available local resources
   - ability for stimulating communal feeling.

Start slowly and on a small scale

This implies starting with the promotion of only a few technologies over small areas and within a micro-catchment basin.

1. This will allow an understanding of the cultural, economic, socio-political, educational, agronomic and ecological factors;
2. it will encourage interchange between the producers and the technicians;
3. it will stimulate creativity;
4. it will reduce the risks and the costs to the farmer;
5. it will allow replying to specific needs (e.g. cultural needs, market needs);
6. it will allow stimulating the interest and involvement of additional farmers.

Experimenting on a small scale

The generation of appropriate technology requires experimentation on a small scale because this:

1. protects the farmer from risks and economic failures;
2. does not require large extra costs;
3. allows the farmer to learn a great deal more;
4. creates a greater willingness to try other innovations;
5. stimulates self-confidence;
6. the technologies will have much greater impact if they are tried out on the farms of the local farmers.

Much more important than the particular technology itself, is the activity of sharing with the community of a method of rural research by which, year after year, they can continue experimenting with innovations. This implies that one is not in the process of developing their agriculture. Rather it is seeking a way as to how they can develop it themselves.

The farmer as a researcher

There are generally few volunteer farmers who use and test the technologies. Testing them allows seeing whether they are adapted to local conditions, whether they are culturally acceptable, or whether they cause environmental damage. The activities and success of farmer volunteers also avoid that the extension worker would appear before the farmers as a simple salesman, only interested in his personal success.

The steps for this methodology are as follows:

1. selection of volunteers;
2. experimentation;
3. joint evaluation of the results by the farmer and the technician;
4. diffusion of the technology.

This methodology can have a slow rate of adoption but over the long term, will be effective.

Avoid technological packages

This implies that one should not bring in a complete and predefined technological package for the farmers, designed on the basis of production objectives. The reasons are described below:

1. the preoccupation of field technicians when working with technological packages is to better understand their details and thus to want to learn methods that will help convince the farmers;
2. generally those who benefit from the technological packages are not the poorest, smallest farmers classified as the target group in the programme objectives;
3. technological packages are practically impossible to apply universally under rural conditions;
4. fundamentally they deal with closed systems that affect the biology of the plant or animal and they do not take account of the social, cultural, organizational, economic and political dimensions of the production work;
5. there is no doubt that farmers need supplies and help from modern society but it is much more appropriate that the farmers themselves take the decisions regarding what is convenient or not within their own strategy and reasoning.

For these reasons, what one can do is to propose a series of alternatives so that the farmer tries out the ideas, adopts them and transforms them.

The role of the technician consists in helping the farmer to understand and to improve the implications of the innovations. It is far preferable to start from the farmer's system, respecting it and enriching it rather than substituting it for a "modern" system enclosed within a package and failing to develop an adapted and sustainable technology.

Possibility of replication

This means the voluntary action of the farmers to adopt and apply a successful technology.

1. it comes from the impact of the successes that emanate from an appropriate technology, which increases and sustains the production and is able to stimulate its own initiatives;
2. it also reaches the poorest people, so allowing them better social justice.

Multiplication is assured when between a quarter and half of the families in a micro-catchment area are successful in applying it. It follows that once this goal has been reached, the extension worker can concentrate on other matters.

Promotional farmers

These are facilitator agents who not only play a lead role in replicating the successful effects but who also assist through their actions and by stimulating creativity in the development process.

Their importance is that within their own cultural context:

- they show greater confidence and so gain greater acceptance for their actions and undertakings;
- they understand better the people with whom they work;
- they understand better the economic problems of the people and of their properties;

- they use a vocabulary that is understood by the people of the community.

The desirable qualities of promotional farmers are as follows:

- they are willing to do what is necessary in order to experiment and to learn;
- they have a constant intention to share with others what they have learnt;
- they set an example in work practices and in innovations;
- they have a constant motivation to learn more;
- they have the capacity to think and to appraise themselves.

Development of self management

This implies following the new tendencies for development that are considered by the local population as the main driving force, with the farmer being the actor rather than the "supplier" of the production process. The farmer is effectively the predominating factor within the ecosystem, conserving it, taking advantage of it or destroying it.

The ability to self-manage, which is understood to mean the efficient management of the resources, implies establishing goals, defining activities and improving the ability to manage both individually and collectively in the solution of local problems.

Apply the technology in stages

This implies that a programme should incorporate only a limited number of innovations. A "staged" technology is one that only changes a few practices in the existing agricultural system. These changes allow the effects to be independently verified as these regard falls or rises that occur in the levels of agricultural production.

Limiting the technology implies that it is "preferable to teach an idea to 100 people rather than 100 ideas to one person".

The reasons for limiting the technology are as follows:

- to start slowly and on a limited scale;
- to guarantee that success will be recognizable;
- to reach the maximum number of people;
- to reach the most critical proportion of the population (at least about a third of the local population);
- to ensure the availability of inputs;
- to promote more social justice;
- to avoid needless effort.

Selection of technologies

In order for a technology to be applied with a harmonious effect, it must comply with the following characteristics:

- it should be well adapted to local conditions;
- it should not involve much investment;
- it should correspond to a need that is felt by the community;
- it should be easy to multiply;
- it should be easy to manage;
- it should produce rapid and noticeable results;
- it should be reasonably permanent;
- it should be easy to organize;
- it should not require too much manual labour.

In reality, it is difficult to find a technology that complies with all these requirements. The selection of a technology will certainly depend on many factors, including on the people involved, the place and the cultural situation.

SOME CRITERIA FOR SELECTING AN APPROPRIATE TECHNOLOGY

1. Does the technology satisfy a need that is felt by the community?
2. Does it bring about economic results?
3. Does it bring about rapid success?
4. Is it applicable to local farming systems?
5. Does it use local resources by preference?
6. Is it easy to understand?
7. Is it low cost?
8. Does it offer low risk?
9. Does it reduce or increase the need for manual labour inputs?

Afterwards, one may compare technologies through a system of points and then decide which technology will be used, as is shown in the following example:

Comparison of technologies

Example 1

Criteria	Scoring		
	Green manure	Chemical fertilizer	Farmyard manure
Ease of application	5	4	2
Multiplier effect	3	4	5
Need felt by the community	2	3	4
Immediate results	5	5	2
Long term effect	5	1	3
Low risk	5	2	3
Cost	5	1	3
Profitability	5	1	3
Ecological appropriateness	5	1	5
Total	40	22	30

In this example, the technology to choose would be green manure. Once the technology has been applied and if the results are positive, then the reaction of the people will probably be positive.

Example 2

Criteria	Minimum tillage	Living barriers	Green manure	Terraces
Use of local resources	5	4	4	3
Low cost	5	5	5	3
Rapid success	5	3	5	4
Ease of understanding	5	5	5	3
Use of manual labour	5	4	4	3
Impact on natural resources	5	5	5	4
Total	30	26	28	20

Sustainability

Sustainability is a characteristic and a criterion for permanent and harmonious development. It implies production without destruction, re-establishing or increasing productivity and conserving the environment.

By using this methodology one achieves one of the two main objectives of sustainable rural development, which is:

"To involve the people in the process so that they gain confidence in themselves and develop innovation as an activity, so giving hope for a better future".

PARTICIPATING WITH THE RURAL FAMILIES IN PLANNING SOIL MANAGEMENT PRACTICES

In the majority of cultures, participation is an art that has to be leant. Constructive participation requires a surprising number of skills.

There is a need for people to learn to:

- express themselves in public;
- analyze and discuss information;
- take decisions;
- resolve differences of opinion;
- criticize constructively with ones companions;
- maintain both horizontal and vertical lines of communication;
- avoid manipulation and influence by autocratic leaders;

It also requires:

- honesty;
- concern;
- mutual trust among the people.

It is important in this process that people know that production increases are feasible, how these increases may most easily be achieved and how this may be taught to others.

At the moment of defining and giving priority to the most important problems, it is fundamental that one avoids the general opinion that "the technician knows everything", but also avoids the trap where the technician would do as little as possible.

Planning of the activities must always be done whilst simultaneously respecting the institutional or technical and scientific approaches and the approaches taken by the community. The institutional approaches rely on basic principles of scientific knowledge whereas the community approaches use their own language based on their own concepts, which are the result of the people living, experiencing and sensing their surroundings.

The system of work or the methodological strategy for implementation of the Proyecto Microcuencas/BIRD in Santa Catarina constantly strives to reconcile these two different approaches. Assistance arising from the technical and scientific approaches starts from the moment of dividing into sectors the existing hydrographic micro-catchment areas within the municipality, by using topographical maps. Once an order of priority has been established for the micro-catchment areas where the work will take place (four during the first six years of Project execution), participatory work starts with the municipal authorities and representatives of the rural communities. This order of priority is established entirely according to technical criteria and following a general diagnostic of the situation of the natural resources in the existing micro-catchment areas.

Work within the priority micro-catchment basin commences with a meeting with the existing community leaders. Later, a meeting is arranged with all the families and the project is presented and discussed. Next, an excursion is made to another micro-catchment basin or to some farms where work has already advanced as regards use, conservation and management of soil and water. The main objective of this trip is to encourage an exchange of experiences among the rural families. This stage is considered important within the work system in order for participatory planning to be achieved for the activities that will be undertaken in the micro-catchment basin. It provides an opportunity for the farmers to learn about the recommended technologies which are already being tried by other families, and the way in which the community has organized itself to solve its problems.

Only after this stage is agreement sought to start the work through signing a document known as the “members’ list”. The group activities and those of a community type which will form part of the management plan for the micro-catchment basin, are established during another meeting with the families. At the same time, responsibilities within the plan are defined for the institutional aspects, for private entities and particularly for the Municipal Prefecture. In addition, a Commission for the Micro-catchment Basin is established democratically, with representatives from the different communities existing within the basin. On this occasion, it is possible to establish a calendar for setting up the conservation plans on the farms of the interested families. Planning for the individual, group and collective activities is done in this way for a period of four years, during which time over the first two years, direct and constant assistance is provided by the extension worker.

COMMUNITY PLANNING

At the present time, the plan for the micro-catchment basin is established during meetings with the communities and on the basis of individual plans for the farms. However, during execution of the second project, which is being formulated to continue with assistance from the World Bank, an effort has been made to improve even more the participation of the rural families. This involves using what are called “Participatory Planning Notebooks” (see Annex 3). With this tool it will be possible to undertake a rapid and integral survey and with greater participation of the rural families. They are very simple and explicit and assist in informally educating the community. This diagnostic allows identification of the main problems existing in the micro-catchment basin, giving them priorities and defining possible corrective actions. The notebooks also serve as a tool for communication between the organized community and the working group from the municipality, in addition to enabling better knowledge to be obtained concerning the reality within the micro-catchment areas.

MAKING THEMATIC MAPS

Thematic maps make up important teaching tools for the participatory planning process. Maps can be used that have been made by the technical and scientific sector as well as those from the community.

Thematic maps from the institutional sector

These are maps made up with the aid of aerial photographs, satellite images, statistical data, etc. In the Proyecto Microcuencas/BIRD developed at Santa Catarina, the field extension workers receive the following tools for assisting them in making the plan for the micro-catchment basin:

1. a hypsometric map;
2. a map of roads and rivers;
3. a map showing actual land use;
4. a map of soil use suitability;
5. a physiographic map;
6. a map of conflicts of use.

The technical and scientific sector can also make other maps, mainly using topographic maps:

1. a map of the hydrographic region;
2. a map of the hydrographic basin;
3. a map of the hydrographic sub-basins;
4. a map of the municipalities within the basin;
5. a map of the micro-catchment basins within the municipality;
6. a map of the micro-catchment basin to be dealt with.

These maps can be needed and can be important during the initial stage of promoting a rural development plan.

Thematic maps from the community

These are made using modern techniques of vision perception. They represent a summary of selective information from the people who live locally. They show a high safety margin and are prepared by extension workers in conjunction with the rural families. These maps are made with assistance from the "Participatory Planning Notebooks" and they help to visualize the data that have been collected through a technique of colouring. A thematic map is prepared for each survey undertaken with a "Participatory Planning Notebook".

Depending upon the objectives of the programme, the following maps may be prepared:

1. a thematic map of the environmental conditions in the micro-catchment basin;
2. a thematic map of the socio-economic conditions;
3. a thematic map of the conditions under which the soils are being managed.

ESTABLISHING PRIORITIES FOR THE ACTIONS TO BE UNDERTAKEN IN THE MICRO-CATCHMENT BASIN

This step in the methodology also constitutes an educational and participatory process whereby the community discusses with the extension worker or the municipal working group, the problems that have been identified in the Integral Diagnostic.

The results of this stage are the proposals that define the main alternatives for solving the identified problems. The order of priority is established through meetings of the community and through municipal seminars.

The order of priorities for the activities defined by the community must appear at the end of the "Participatory Planning Notebooks" and must be established so as to reply to the following questions:

- What must we do to solve our problems?
- How can we solve our problems?
- Who can help us to solve our problems?
- When can we solve our problems?

The work plan to be formulated is based on the structure and the calendar of the activities to be undertaken with the farmers from the micro-catchment basin. It is also based on the costs and the necessary infrastructure and it identifies the levels of responsibility of the individuals, groups or the institutions involved.

FORMULATION OF PROJECTS

Often, because of the type of activity that has been given priority by the rural families, it is necessary to formulate specific projects for requesting external aid.

Within the present Proyecto Microcuencas/BIRD developed in Santa Catarina for example, there is a component that provides resources for improving and relocating the community roads. However, in order for the community to define democratically the distance to be improved, execution of this activity necessarily requires the assistance of an engineer, mainly for elaborating the project. Furthermore, the project will be executed by the private sector, selected through a bidding process.

Municipal Prefectures, Government Institutions and Non-Government Organizations can also formulate projects to complement the activities to be undertaken with the rural communities in the selected micro-catchment basin.

However, it should be pointed out that the projects must be formulated according to the socio-environmental and economic dimensions of the local population and generally, these are as follows:

- Community projects;
- Group projects;
- Individual projects.

CONSERVATION FARM PLANNING

The Individual Farm Plan (IFP)

The methodology developed for conservation farm planning which is being used in Santa Catarina, is based on soil use according to its natural suitability as determined by criteria established for the hilly conditions that predominate in the region. The objective of the IFP is to implement a participatory proposal for improving the particular rural property through a timetable of activities over a four-year period.

Formulation of the IFP can be divided into various steps as follows:

1. Identification of the farmer

2. Survey and diagnostic of the farm:
   - the situation of the annual crops and other activities
   - the use of conservation practices
   - animal stocks
   - environmental sanitation

3. Drawing a map of Actual Land Use and Land Use Suitability.

4. Preparation of a table showing the land distribution.

5. Conservation planning and recommended practices:
   - preparation of a table showing the redistribution of the land use
   - preparation of a map showing the redistribution of the land use
   - land use and land management
   - environmental improvements
   - timetable for the implementation of the improvements and investments.

COMMUNITY MAPPING

Experience gained in other countries shows that it is possible to make community maps, which indicate how the land is being used in the micro-catchment area according to its natural suitability, even though the majority of the local population may be illiterate. It is often surprising to the technicians, how the rural families are able to establish detailed maps showing the situation in the micro-catchment area. These are often made using symbols or whatever other items related to nature that are available such as seeds, wood, fruit or stone and the maps can often include the exact location of those areas with problems of deterioration of the natural resources. In this case, it is essential to work in the open air, often using the very same soil for drawing the maps of the micro-catchment basins. When there are some people within the community who know how to write and to draw, it is possible to carry out this activity in a classroom, for example using paper to make the maps. It is also possible to make maps for redistributing the use of land that is presently being used beyond its natural suitability, through an in-depth discussion between the community families and assisted by the extension worker, which examines the areas with problems.

IMPLEMENTATION OF SOIL MANAGEMENT PLANS

The implementation of a plan is one of the most fascinating and significant phases of the participatory planning process because starting from this, fundamental modifications commence for the future of the community. During this stage, all the practical skills and knowledge of residents within the micro-catchment basin are used to legalize and execute activities or works within the community, the fruit of their participatory effort.

Implementation of the work is fundamental for the micro-catchment basin plan, which is the end-result of the discussion between the local population and the extension worker or the working group from the municipality. During the process of execution, it is essential that every institution and every family involved in the micro-catchment basin take up their responsibilities according to what has been established in the plan.

The technical assistance input must be introduced through an educational and participatory process (meetings, seminars, demonstrations, field visits). Field visits should be arranged during the process of formulating the Individual Plans and during the application of resources for the Programme of Incentives for Introducing Soil Conservation Practices.

It is fundamental to establish a timetable clearly indicating the actions to be undertaken within the micro-catchment basin. In reality, this plan is no more than the formulation of a "Plan of Operations", which describes a methodological strategy for identifying how the activities will be implemented in the micro-catchment basin.

For this, it is necessary to specify how to undertake the following:

1. How to generate initial motivation amongst the families regarding the technologies for soil management:
   - arrange meetings for generating motivation?
   - arrange field trips to learn about other successful experiences?
   - use audio-visual aids to show the processes of soil degradation and how soil recovery may be accomplished?

2. Implement the technological proposals within the micro-catchment basin:
   - which technologies should be selected?
   - are demonstration areas needed?
   - how will the results be demonstrated?

- can feedback be arranged for the trials?
- Should small-scale trials be established first?

3. Make technical assistance available:
   - to groups of neighbours?
   - to individuals?
   - through an exchange of experiences?
   - through training exercises?

4. Implement the soil management practices:
   - should machinery from hire-pools or communally owned equipment be used?
   - should machinery be used that is destined for community use?
   - should individually owned machinery be used?

5. Mobilize the community to implement the activities included in the plan:
   - what will be the role of the Commission for the Micro-catchment Basin?

In summary, the "Plan of Operations" should be formulated in such a way as to answer the questions: "What should be done?" "How will it be done?" "When will it be done?" and "Who will do it?".

It is most important that the Plan of Operations is filled out completely, and this should then stay together with the other part of the "Participatory Planning Notebook".

Bibliography

Adams, J.E. 1966. Influence of mulches on runoff, erosion and soil moisture depletion. Soil Sci. Soc. Am. Proc., *Madison*. 30: 110-14.

Almeida, F.S. 1985a. Efeitos alelopáticos da cobertura morta. Plantio direto, Ponta Grossa, III (10): 4-5.

Almeida, F.S. 1985b. Influência da cobertura morta do plantio direto na biología do solo. In: *Atualização em plantio direto*. Fancelli, A.L.; Tonado, P.V. & Machado, J. Campinas, Fundação Cargill. Chapter 6, pp.103-144.

Almeida, F.S. 1987. Alelopatia das coberturas mortas. Plantio direto, Ponta Grossa, V(21): 4-5.

Almeida, F.S.; Rodrigues, B.N. and Oliveira, V.F. 1984. Plantio Direto de Milho: Uso de coberturas mortas como forma de reduzir o emprego de herbicidas. In: *Congresso nacional de milho e sorgo*, **15**, Maceió, 1984. Resumos... Sete Lagoas, EMBRAPA/CNPMS. p.72.

Almeida, R.A. 1993. Adaptação da " matraca " ao plantio direto em pequenas propriedades. In: *Encontro latinoamericano sobre plantio direito na pequena propriedade*, **1**, Ponta Grossa, PR. Anais. SEAB-IAPAR, Ponta Grossa. pp. 251-257.

Alves, C.S. 1978. Controle e estabilização de voçorocas. Trigo e Soja - *Boletim Técnico FECOTRIGO*, **37**, Porto Alegre. pp. 6-9.

Amado, T.J.C. 1985. Relações da erosão hidrica dos solos com doses e formas de manejo do residuo da cultura da soja. Porto Alegre. Thesis (Mestrado Agronomía). Fac. Agronomía, Universidade Federal do Río Grande do Sul, Porto Alegre. 104 p.

Amado, T.J.C. 1991. Adubação verde de inverno para o Alto Vale do Itajaí. Rev. Agropec. Catarinense, Florianópolis, 4 (1): 4-7.

Amado, T.J.C. and Wildner, L. do P. 1991. Adubação verde. In: *Manual de Uso, Manejo e Conservação do Solo e da Água*. Santa Catarina. Secretaria de Estado da Agricultura e Abastecimento. Florianópolis. pp. 105-117.

Amado, T.J.C.; Matos, A.T. de, and Torres, L. 1990. Flutuação de temperatura e umidade do solo sob preparo convencional e em faixas na cultura da cebola. Pesq. Agropec. Bras., Brasília, 25(4): 625-31.

Anjos, J.B. dos, Soares, J.G.G., and Baron, V. 1983. Adaptação de plantadeira manual para o plantio de sementes de capim búfel. EMBRAPA - CPATSA, Petrolina, PE. (EMBRAPA-CPATSA. *Documentos,* **21**). 5 p.

Anjos, J.B. dos. 1985. Comparação entre tração motorizada e animal. In: *Simpósio sobre energia a agricultura, tecnologias poupadoras de insumos, integração de sistemas energéticos de alimentos,* 1., 1984, Jaboticabal-SP. Anais. Jaboticabal: FCAV. pp. 269-289.

Anjos, J.B. dos., Baron, V. and Bertaux, S. 1988. Captação de água de chuva "in situ " com aração parcial. EMBRAPA-CPATSA, Petrolina, PE. (EMBRAPA-CPATSA. *Comunicado Técnico,* **26**). 4 p.

Archer, J.R. and Smith, P.D. 1972. The relation between bulk density, available water capacity and air capacity of soil. *Journal of Soil Science*, London, **23**:475-80.

Argueta, M.T. 1996. Analysis of the production and utilization of crop residues and their effect on erosion risk in the department of Morazan. Agronomy Engineer's thesis, Faculty of Engineering, Central American University "José Simeón Canas," San Salvador, El Salvador.

Argueta, M.T. 1996. Análisis de la producción y utilización de rastrojos y su efecto sobre el riesgo de erosión en el departamento de Morazan. Tesis de Ingeniero Agrónomo, Facultad de Ingeniería, Universidad Centroamericana "José Simeón Canas," San Salvador, El Salvador.

Barber, R.G. 1994. Rotaciones de cultivos para zonas con 1000 a 1300 mm de lluvia por año en el departamento de Santa Cruz, Bolivia. Centro de Investigación Agrícola Tropical, Misión Británica en Agricultura Tropical, y Banco Mundial y Proyecto de Tierras Bajas del Este, Santa Cruz, Bolivia.

Barber, R.G. 1998. Linking the production and use of dry-season fodder to improved soil conservation practices in El Salvador. *Proceedings of the 9th. International Soil Conservation Organization Conference*, Bonn, Germany, 26-30 August, 1996. Advances in GeoEcology 31, Volume II 1311-1317. Catena-verlag, Reiskirchen, Germany.

Barber, R.G. and Díaz, O. 1992. Effects of deep tillage and fertilization on soya yields in a compacted Ustochrept during seven cropping seasons, Santa Cruz, Bolivia. *Soil and Tillage Research*, **22**, 371-381.

Barber, R.G. and Johnson, J. 1992. Aspectos técnicos sobre la instalación de cortinas rompevientos para la producción de cultivos anuales. *Informe Técnico No.* **2**. Centro de Investigación Agrícola Tropical y Misión Británica en Agricultura Tropical, Santa Cruz, Bolivia.

Barber, R.G. and Thomas, D.B. 1981. Infiltration, surface runoff and soil loss from high intensity simulated rainfall in Kenya. Research contract No. RPIHQ 1977-3/AGL, FAO, Rome. Italy. Faculty of Agriculture, University of Nairobi, Nairobi, Kenya.

Barber, R.G., Herrera, C. and Díaz, 0. 1989. Compaction status and compaction susceptibility of alluvial soils in Santa Cruz, Bolivia. *Soil and Tillage Research*, **15**, 153-167.

Barber, R.G., Navarro, F. and Orellana, M. 1993. Labranza Vertical. Centro de Investigación Agrícola Tropical, Misión Británica en Agricultura Tropical y Proyecto de Desarrollo Tierras Bajas del Este del Banco Mundial, Santa Cruz, Bolivia.

Barbosa, L.R., Díaz, O. and Barber, R.G. 1989. Effects of deep tillage on soil properties, growth and yield of soya in a compacted Ustochrept , Santa Cruz, Bolivia. *Soil and Tillage Research*, **15**, 51-63.

Beek, K.J. 1975. Recursos naturais e estudos perspectivos a longo prazo: notas metodológicas. Brasília, SUPLAN. 57 p.

Bennema, J.; Beek, K.J.; and Camargo, M.N. 1964. Interpretação de levantamento de solos no Brasil. Rio de Janeiro, Ministério da Agricultura. 50 p.

Bernardo, S. 1982. Manual de irrigação. Viçosa: UFV. 463 p.

Bertolini, D., Galetti, P.A., and Drugowich, M.I. 1989. Tipos e formas de terraços. In: *Simpósio sobre terraceamento agrícola*, 1988, Campinas, SP. Anais. Fundação Cargill, Campinas. pp. 79-98.

Bertoni, J. and Lombardi Neto, F. 1985. Conservação do solo. Livroceres, Piracicaba. 392 p.

Bertoni, J.; Pastana, F.I.; Lombardi Neto, F. and Benatti Júnior, R. 1972. Conclusões gerais das pesquisas sobre conservação do solo no Instituto Agronômico. (*IAC, Circular No.* **20**) Instituto Agronômico, Campinas. 56 p.

Borst, H.L. and Woodburn, R. 1942. The effect of mulching and methods of cultivation on runoff and erosion from Muskingum silt loam. Agr. Engin., 23(1): 19-22.

Brady, N.C. 1974. Nature and Properties of Soils. 8th. ed. New York, McMillan. 639p.

Bulisani, E.A. and Roston, A.J. 1993. Leguminosas: adubação verde e rotação de culturas. In: *Curso sobre Adubação Verde no Instituto Agronômico*, **1**. Wutke, E.B.; Bulisani, E.A. & Mascarenhas, H.A.A. Campinas: Instituto Agronômico. pp. 13-16. (Documentos IAC 35).

Cabeda, M.S.V. 1984. Degradação física e erosão. In: *I Simpósio de manejo do solo e plantio direito no sul do Brasil e III Simpósio de conservação de solos do planalto*. Passo Fundo, RS, 1983. Anais.

Calegari, A. 1989. Cobertura morta. In: Secretaria da Agricultura e do Abastecimento. Paraná. *Manual Técnico do Subprograma de Manejo e Conservação do Solo.* Curitiba. pp. 212-17.

Cannell, R.Q. and Finney, J.R. 1973. Effects of direct drilling and reduced cultivation on soil conditions for root growth. Outlook on Agriculture, Bracknell, 7:184-9.

Carvalho, O. de. (coord). 1973. Comportamento dos principais sistemas de produção da zona semiárida. In: *Plano integrado para o combate preventivo aos efeitos das secas no Nordeste*, Brasília: MINTER. (Brasil. MINTER. Desenvolvimento Regional. Monografias, 1). pp. 105-119.

Castro Filho, C. 1989. Cordões em contorno e cordões de vegetação permanente. In: *Manual Técnico do Sub-programa de manejo e conservação do solo.* Curitiba. Secretaria da Agricultura e do Abastecimento. Paraná. pp. 236-238.

Cintra, F.L.D. 1980. Caracterização do impedimento mecânico em Latossolos do Rio Grande do Sul. 89p. Faculdade de Agronomia, UFRGS, Porto Alegre, 1980.

Cochrane, T.T. and Barber, R.G. 1993. Análisis de Suelos y Plantas Tropicales. Centro de Investigación Agrícola Tropical y Misión Británica en Agricultura Tropical, Santa Cruz, Bolivia.

Cook, M.G. and Lewis, W.M. (eds.) 1989. Conservation tillage for crop production in North Carolina. North Carolina Agricultural Extension Service, North Carolina State University, Raleigh, N.C. USA.

Costa, M.B.B. da (Coord.); Calegari, A.; Mondardo, A.; Bulisani, E.A.; Wildner, L. do P.; Alcântara, P.B.; Miyasaka, S. and Amado, T.J.C. 1992. Adubação verde no sul do Brasil. Rio de Janeiro, AS-PTA. 346 p.

Curi, N. (Coord.). 1993. In collaboration with J.O.I. Larach, N. Kampf, A.C. Moniz and L.E.F. Fontes. Vocabulário de ciência do solo. Campinas, Sociedade Brasileira de Ciência do Solo. 90 p.

Daker, A. 1988. Irrigação e drenagem: a água na agricultura, 7 ed. rev. e ampl. Rio de Janeiro: F. Bastos, v.3. 543 p. ill.

Daniel, L.A. 1989. Canais escoadouros e dissipadores. In: Simpósio sobre terraceamento agrícola, 1988, Campinas, SP. Anais. Fundação Cargill, Campinas. pp.233-255.

Dent, D. and Young, A. 1981. Soil survey and land evaluation. Norwich, School of Environmental Sciences, University of East Anglia. 278 p.

Derpsch, R. 1984. Alguns resultados sobre adubação verde no Paraná. In: *Adubação verde no Brasil.* Fundação Cargill, Campinas. pp. 268-79.

Derpsch, R.; Sidiras, N. and Heinzmann, F.X. 1985. Manejo do solo com coberturas verdes de inverno. *Pesq. Agropec. Bras.* Brasília, **20** (7): 761-73.

Duque, J.G. 1973. Algumas questões da exploração de açudes públicos. In: *Solo e água no polígono das secas*. **4**. ed. Fortaleza : DNOCS, 1973. pp. 129-156. (DNOCS. Publicação,- 154).

Duret, T.; Baron, V.; and Anjos, J.B. dos. 1986. "Systèmes de cultures" expérimentés dans le Nordest du Bresil. *Machinisme Agricole Tropicale, Antony*, **94,** pp. 62-74.

Edwards, J.H., Wood, C.W., Thurlow, D.L. and Ruf, M.E. 1992. Tillage and crop rotation effects on fertility status of a Hapludult soil. *Soil Science Society of America Journal*, **56**, 1577-1582.

EMBRAPA (Centro de Pesquisa Agropecuária do Trópico Semi-Árido). 1988. Irrigação por pivô central no Serviço de Produção de Sementes Básicas (Bebedouro II): avaliação técnico-econômica. Petrolina, PE. (EMBRAPA-CPATSA. *Documentos*, **51**) 100 p. EMBRAPA-CNPT, Passo Fundo. pp. 139-141.

Erbach, D.C. 1994. Benefits of Tracked Vehicles in Crop Production. In: *Soil Compaction in Crop Production*. B.D. Soane and C. van Ouwerkerk (Eds.), Amsterdam.

FAO. 1979. Yield response to water. J. Doorenbos and A.H. Kassam. FAO Irrigation and Drainage Paper **33**. Rome, FAO.

Ferreira, L.A.B. 1984. Vegetaçáo para a fixação de taludes. Trigo e Soja - *Boletim Técnico FECOTRIGO*, **74**, Porto Alegre. pp. 18-21.

Finkel, H.J. 1986. Wind erosion. pp. 109-121. In: *Semiarid Soil And Water Conservation* H.J. Finkel, M. Finkel and Zeév Naveh (eds.), CRC Press Inc. Boca Raton, Fl. USA.

França da Silva, I. 1980. Efeitos de sistemas de manejos e tempo de cultivo sobre as propriedades físicas de um Latossolo. 70f. Thesis (Mestr. Agron. - Solos) Faculdade de Agronomia, UFRGS, Porto Alegre, 1980. (unpublished).

Franz, C.A.B.; and Alonço, A.S. 1986. Sulcador acoplável a semeadeiras-adubadeiras para a implantação de lavouras irrigadas por sulcos. EMBRAPA CPAC, Planaltina, DF. (EMBRAPA-CPAC. *Circular Técnico*, **24**). 22 p. il.

Gogerty, R. 1995. When One Tillage System Isn't Enough. *The Furrow* **100**(4): 37-38.

Guerra, P. de B. 1975. Agricultura de vazante - um modelo agronômico nordestino. In: *Seminário nacional de irrigação e drenagem*, **3**., Fortaleza, CE. Anais. Recife: DNOCS/ABID, 1976. v. 4, pp. 325-33.

Hull, W.X. 1959. Manual de conservação do solo. Secretaria da Agricultura dos Estados Unidos da América, Washington D.C. (Publicação TC-284) pp. 71-83.

IAPAR (Fundação Instituto Agronômico do Paraná). 1978. Relatório Técnico Anual 1977 - Programa de Manejo e Conservação de Solos. Londrina, PR. pp. 221-231.

IAPAR (Fundação Instituto Agronômico do Paraná). 1984. IAPAR 10 anos de pesquisa: relatório técnico 1972-1982. Londrina, PR, Relatório Técnico. 233 p.

Kayombo, B. and Lal, R. 1994. Responses of Tropical Crops to Soil Compaction. In: *Soil Compaction in Crop Production*, B.D. Soane and C. van Ouwerkerk (Eds.). Amsterdam.

Kepner, R.A., Bainer, R. and Barger, E.L. 1972. Chisel-type and multipowered tillage implements. In: *Principles of machinery*. 2. ed., Kepner, R.A.; Bainer, R.; Barger, E.L. Westport: AVI. 203 p.

Kitamura, P.C. 1982. Agricultura migratória na Amazonia: un sistema viável? EMBRAPA-CPATU, Belém, PA. (EMBRAPA-CPATU. *Documentos*, **12**). 20 p. ill.

Klamt, E. 1978. Avaliação dos sistemas de classificação da capacidade de uso dos solos. In: *II Encontro nacional de pesquisa sobre conservação do solo,* **2**, 1978, Passo Fundo. Anais. Passo Fundo, EMBRAPA, CNPT. pp. 453-463.

Klamt, E. and Stammel, J.G. 1984. Manejo adequado dos solos das encostas basálticas. Porto Alegre, *Trigo e Soja,* **74**, pp. 4-11, July/August.

Klingebiel, A.A. and Montgomery, P.H. 1961. Land capability classification. *Agriculture Handbook,* **210**. Washington, USDA. 21 p.

Kohnke, H. 1968. Soil physics. McGraw Hill, New York. 224 p.

Kohnke, N. and Bertrand, A.R. 1959. Soil Conservation. McGraw-Hill, New York. 298 p.

Lal, R. 1975. Role of mulching techniques in tropical soil and water management. *IITA Technical Bulletin,* **1**. IITA, Ibadan, Nigeria.

Lal, R. 1985. A soil suitability guide for different tillage systems in the tropics. *Soil and Tillage Research,* **5**: 179-196.

Lal, R. 1995. Tillage systems in the tropics: Management options and sustainability implications. *Soils Bulletin* **71**. FAO, Rome, Italy.

Lal, R., De Vleeschauwer, D. and Malafa Nganje, R. 1980. Changes in properties of a newly cleared tropical Alfisol as affected by mulching. *Soil Science Society Am J.* **44**: 827-833.

Langdale, G.W., West, L.T., Bruce, R.R., Miller, W.P. and Thomas, A.W. 1992. Restoration of eroded soil with conservation tillage. Soil Technology **5**: 81-90.

Larson, W.E. 1964. Soil parameters for evaluating tillage needs and operations. *Soil Science Society of America Proceedings, Madison.* 28:118-22.

Larson, W.E.; Eynard, A.; Hadas, A. and Lipiec, L. 1994. Control and Avoidance of Soil Compaction in Practice. In: *Soil Compaction in Crop Production*, B.D. Soane and C. Van Ouwerkerk (Eds.). Amsterdam.

Lattanzi, A.R.; Meyer, L.D. and Baumgarner, M.F. 1974. Influences of mulch rate and slope steepness on internal erosion. *Soil Sci. Soc. Am. Proc., Madison.* 38:946-50.

Lepsch, I.F. (Coord.) 1983. Manual para levantamento utilitário do meio físico e classificaçáo de terras no sistema de capacidade de uso. Campinas, Sociedade Brasileira de Ciência do Solo. 175 p.

Lepsch, I.F. (Coord.). 1991. Manual para levantamento utilitário do meio físico e classificação de terras no sistema de capacidade de uso. 4a aprox. Campinas, Sociedade Brasileira de Ciência do Solo. 175 p.

Lombardi Neto, F.; Bellinazzi Júnior, R.; Lepsch, I. F.; Oliveira, J.B. de; Bertolini, D.; Galeti, P. A. and Drugowich, M. I. 1991. Terraceamento Agrícola. Campinas, SP, Brasil, Coordenadoria de Assistência Técnica Integral. (Boletim Técnico, **206**). 38p.

Lopes, P.R.C. 1984. Relações da erosão com tipos e quantidades de resíduos culturais espalhados uniformemente sobre o solo. Dissertation (Mestrado Agronomia). Fac. Agronomia, Universidade Federal do Rio Grande do Sul. Porto Alegre. 116 p.

Lopes, P.R.C., and Brito, L.T. de L. 1993. Erosividade da chuva no Médio São Francisco. *Revista brasileira de Ciência do Solo,* Campinas, v. **17**, n.l, pp. 129-133.

Lorenzi, H. 1984. Inibição alelopática de plantas daninhas. In: *Adubação verde no Brasil.* Fundação Cargill, Cap. 20, Campinas. p. 1983-98.

Lowry, F.E., Taylor, H.M., and Huck, M.G. 1970. Root elongation rate and yield of cotton as influenced by depth and bulk density of soil pans. *Soil Science Society America Proceeding, Madison,* **34**:306-9.

Mannering, J.V. and Meyer, L.D. 1963. The effect of various rates of surface mulch on infiltration and erosion. *Soil Sci.Soc.Am. Proc., Madison,* **27**:84-6.

Marques, J.Q.A. 1971. Manual brasileiro para levantamento da capacidade de uso da terra. 3ª aprox. Rio de Janeiro, IBGE. 433 p.

Mazuchowski, J.Z., and Derpsch, R. 1984. Guía de preparo do solo para culturas anuais mecanizadas. ACARPA, Curitiba. 65 p. ii.

Mejía, M. 1984. Nombres Científicos y Vulgares de Especies Forrajeras Tropicales. Centro Internacional de Agricultura Tropical (CIAT), Cali, Colombia. 75p.

Meyer, L.D.; Wischmeier, W.H. and Foster, G.R. 1970. Mulch rates required for erosion control on steep slopes. *Soil. Sci. Soc. Am. Proc., Madison,* **34**:928-31.

Mielniczuk, J. and Schneider, P. 1984. Aspectos sócioeconômicos do manejo de solos no sul do Brasil. In: *I Simpósio de manejo do solo e plantio direto no sul do Brasil e III Simpósio de conservação de solos do planalto.* Passo Fundo, RS, 1983. Anais.

Miyasaka, S. 1984. Histórico de estudos de adubação verde, leguminosas viáveis e suas características. In: *Encontro nacional sobre adubaçáo verde,* 1, 1983, Rio de Janeiro. Adubação verde no Brasil. Fundação Cargill, Campinas. pp. 64-123.

Monegat, C. 1981. A ervilhaca e o cultivo mínimo. ACARESC, Chapecó. 24p.

Monegat, C. 1991. Plantas de cobertura do solo: características e manejo em pequenas propriedades. Ed. do autor, Chapecó. 337 p.

Mullins, C.E., MacLeod, D.A., Northcote, K.H., Tisdall, J.M. and Young, I.M. 1990. Hardsetting soils: Behaviour, Occurrence and Management. *Adv. Soil Sci.* **11**, 37-108.

Musgrave, G.W. and Nichols, M. L. 1942. Organic matter in relation to land use. *Soil Sci. Soc. Am. Proc., Madison.* 7:22-28.

Muzilli, O.; Oliveira, E.E.; Gerage, A.C. and Tornero, M.T. 1983. Adubação nitrogenada em milho no Paraná: III. Influência da recuperação do solo com adubação verde de inverno na resposta à adubação nitrogenada. Pesq. Agropec. Bras., Brasília, 18 (1): 23-27.

Muzilli, O.; Vieira, M.J. and Parra, M.S. 1980. Adubação verde. In: *Manual Agropecuário para o Paraná,* Chapter 3, Fundação Instituto Agronômico do Paraná. pp. 76-93.

Naderman, G.C. Jr 1990. An overview: Subsurface compaction and subsoiling in North Carolina. North Carolina Agricultural Extension Service, North Carolina State University, Raleigh, N.C. USA.

Nangju, D., Wien, H.C. and Singh, T.P. 1975. Some factors affecting soybean viability and emergence in the lowland tropics. *Proc. Soybean Research Conference*, Urbana, 3-8 August 1975.

Orellana, M., Barber, R.G. and Díaz, O. 1990. Effects of deep tillage and fertilization on the population, growth and yield of soya during exceptionally wet season on a compacted sandy loam, Santa Cruz, Bolivia. *Soil and Tillage Research,* **17**: 47-61.

Paningbatan, E.P. Ciesiolka, C.A., Coughlan, K.J. and Rose, C.W. 1995. Alley cropping for managing soil erosion of hilly lands in the Phillipines. Soil Technology, 8. 193-204.

Prentice, A.N. 1946. Tied-ridging with special reference to semi-arid aereas. East African J. **12**: 101-108.

Pundek, M. O. 1985. Manual de Conservação do Solo. ACARESC, Florianópolis. 23 p.

Ramallo Filho. A. and Beek, K.J. 1995. Sistema de avaliaçáo da aptidão agrícola das terras. 3 ed. Brasília, EMBRAPA, CNPS. 65 p.

Ramalho Filho, A., Pereira, E.G., and Beek, K.J. 1977. Sistema de avaliação da aptidão agrícola do solo. Brasília, SUPLAN. 26 p.

Ramalho Filho, A., Pereira, E.G. and Beek, K.J. 1978. Sistema de avaliaçáo da aptidão agrícola das terras. Brasília, EMBRAPA, SNLCS. 70 p.

Resende, M., Curi, N., Rezende, S.B., and Corréa, G.F. 1995. Pedologia: base para distinção de ambientes. Viçosa, NEPUT. 304 p.

Reynolds, R. 1995. New Ways to Break Old Ground; Deere & Company, Moline/Illinois, USA.

Rice, R.W. (ed.) 1983. Fundamentals of no-till farming. American Association for Vocational Instructional Materials, Athens, GA. USA.

Rio Grande do Sul. 1985. Secretaria da Agricultura. Manual de Conservação do Solo. 3rd updated ed. Porto Alegre. 287 p.

Rockwood, W.G. and Lal, R. 1974. Mulch Tillage: A technique for soil and water conservation in the tropics. SPAN 17: 77-79.

Rowell, D.L. 1994. Soil Science: methods and applications. London. Longman. 350 p.

Rufino, R.L. 1989. Terraceamento. In: *Manual Técnico do Subprograma de Manejo e Conservação do Solo,* Curitiba. Secretaria da Agricultura e do Abastecimento, Paraná. pp. 218-235.

Santos, H.P. dos; Reis, E.M. and Pottker, D. 1987. Efeito da rotação de culturas no rendimento de grãos e na ocorrência de doenças radiculares de trigo (*Triticum aestivum*) e de outras culturas de inverno e de verão de 1979 a 1986. Passo Fundo: EMBRAPA-CNPT, (*Documentos*, 7). 38 p.

Scherer, E.E. and Baldissera, I.T. 1988. Mucuna: a proteção do solo em lavoura de milho. Rev. Agrop. Catarinense, Florianópolis, 1(1): 21-25.

Sharma, B.R. 1991. Effect of different tillage practices, mulch and nitrogen on soil properties, growth and yield of fodder maize. *Soil and Tillage Research,* **19**: 55-66.

Sharma, R.D.; Pereira, J. and Resck, D.V.S. 1982. Eficiência de adubos verdes no controle de nematódes associados a soja nos cerrados. Planaltina: EMBRAPA/CPAC, *(Boletim de Pesquisa,* **13**) 30p.

Shaxson T.F., Hudson, N.W., Sanders, D.W., Roose, E. and Moldenhauer, W.C. 1989. Land husbandry: a framework for soil and water conservation. Soil and Water Conservation Society and the World Association of Soil and Water Conservation, Ankeny, Iowa, USA.

Sidiras, N. and Roth, C.H. 1984. Medições de infiltração com infiltrômetros e em simulador de chuvas em Latossolo Roxo distrófico, Paraná, sob vários tipos de cobertura do solo e sistema de preparo.

In: *Congresso Brasileiro de conservação do solo*, **5**. Porto Alegre, RS. 1984. Summary. Porto Alegre.

Silva, A. de S., and Porto, E.R. 1982. Utilização e conservação de recursos hídricos em áreas rurais do Trópico Semi-árido do Brasil: tecnologias de baixo custo. Petrolina, PE. EMBRAPA-CPATSA. (EMBRAPA-CPATSA. *Documentos*, **14**). 128 p. il.

Silva, D.A., Silva, A.S., and Gheyi, H.R. 1981. Pequena irrigação para o trópico semi-árido: Vazantes e cápsulas porosas. Petrolina, PE: EMBRAPA-CPATSA. (EMBRAPA - CPATSA. *Boletim de Pesquisa*, **3**). 59 p. il.

Silva, E.; Teixeira, L.A.J. and Amado, T.J.C. 1993a. Kit de microtrator para cultivo mínimo da cebola. In: *Encontro Latinoamericano sobre plantio direto na pequena propriedade*, **1**, Ponta Grossa. Anais. IAPAR, Ponta Grossa. pp. 265-270.

Silva, F.B.R.E., Riché, G.R., Tonneau, J.P., Souza Neto, N.C. de, Brito, L.T. de L., Correia, R. C., Cavalcanti, A.C., Silva, F.H.B.B. da, Silva, A.B. da, and Araújo Filho, J.C. de. 1993b. Zoneamento agroecológico do Nordeste: diagnóstico do quadro natural e agrossócioeconômico. Petrolina, PE: EMBRAPA - CPATSA/Recife: EMBRAPA - CNPS, Coordenadoria Regional Nordeste. v. 1 il.

Singer, M.J. Matsuda, Y. and Blackard, J. 1981. Effect of mulch rate on soil loss by raindrop splash. *Soil Sci. Soc. Am. Journal, Madison.* **45**: 101-110.

Singer, M.J. and Blackard, J. 1978. Effect of mulching on sediment in runoff from simulated rainfall. Soil Sci. Soc. Am. Proc., Madison. 42:481-86.

Sloneker, L.L. and Moldenhauer, W.C. 1977. Measuring the amounts of crop residue remaining after tillage. *Journal Soil and Water Conservation, Ankeny, Iowa*, **32**(5): 231-236.

Soares, J.M. 1988. Sistema de irrigação por inundaçâo. Petrolina, PE: EMBRAPA-CPATSA, (EMBRAPA-CPATSA, *Documentos*, **55**). 49 p. il.

Sobral Filho, R.M., Madeira Neto, J. Das, Freitas, P.L. de, and Silva, R.L.P. da. 1980. Práticas de Conservação de Solos. (EMBRAPA-SNCLS, Miscelânea, 3). EMBRAPA-SNCLS, Rio de Janeiro. 88 p.

Steiner, J.L., Schomberg, H.H. and Morrison, J.E. 1994. Residue decomposition and redistribution. In: Crop residue management in the southern Great Plains. U.S. Dept. Agric., Agric. Res. Serv. Info. Staff, Beltsville, MD, USA.

Streck, E.V. 1992. Levantamento de solos e avaliação do potencial agrícola das terras da microbacia do Lageado Atafona (Santo Ângelol/RS). Porto Alegre, Faculdade de Agronomia, UFRGS. 167p. Master's thesis in Agronomia - Ciência do Solo.

Taylor, J.H. 1994. Development and Benefits of Vehicle Gantries and Controlled Traffic Systems. In: Soil Compaction in Crop Produdion, B.D. Soane and C. van Ouwerkerk (Eds.). Amsterdam.

Tuler, V.V.; Nascif, A.F.; Souza, D. de; Azevedo, H.J. de. 1983. Controle da irrigação pelo tanque classe A. Piracicaba: Programa Nacional de Melhoramento de Cana-de Açúcar. 12p.

Uberti, A.A.A. 1985. Alternativas para uso adequado dos solos do Oeste e Vale do Rio do Peixe. Florianópolis, EMPASC. *Comunicado Técnico*, **85**. 7 p.

Unger, P.W., Jones, O.R. and Laryea, K.B. 1995. Sistemas de labranza y prácticas de manejo de suelos para diferentes condiciones de tierras y climas. In: *Memorias de la segunda reunión bienal de la*

Red Latinoamericana de Labranza Conservacionista, Eds. I. Pla Sentís and F. Ovalles, Guanare, Acarigua, Venezuela, RELACO. pp. 82-117.

Valdiviezo Salazar, C.R.; Cordeiro, G.G. 1985. Perspectivas do uso das águas subterrâneas do embasamento cristalino no Nordeste semi-árido do Brasil. Petrolina, PE: EMBRAPA - CPATSA. (EMBRAPA - CPATSA. *Documentos*, **39**). 40 p. il.

Vermeulen, G.D. and Perdok, U.D. 1994. Benefits of Low Ground Pressure Tyre Equipment. In: Soil Compaction in Crop Production, B.D. Soane and C. van Ouwerkerk (Eds.). Amsterdam.

Vieira, M.J. 1987. Solos de baixa aptidão: opções de uso e técnicas de manejo e conservação. (IAPAR, *Circular No.* **51**). IAPAR, Londrina. 68 p.

Voorhees, W.B., Senst, C.G., and Nelson, W.W. 1978. Compaction and soil structure modification by wheel traffic in the Northern Corn Belt. *Soil Science Society of America Journal, Madison* **42**:344-9.

Wieneke, F. and Th. Friedrich, 1989: Agricultural Engineering in the Tropics and Subtropics; Centaurus Verlag, Pfaffenweiler.

Wildner, L. do P. 1990. Adubação verde, cobertura e recuperação do solo em sistemas diversificados de produção. Chapecó, EMPASC/CPPP, 1986. 79 p. (EMBRAPA, PNP Manejo e Conservação do Solo. Projeto 04386007/1). Relatório final.

Wildner, L. do P. and Massignam, A.M. 1994a. Ecofisiologia de alguns adubos verdes de verão: I. Produção de fitomassa - resultados preliminares. In: *Reunião Centro-sul de adubação verde e rotação de culturas*, **4**, 1993, Passo Fundo, RS. Anais. (EMBRAPA - CNPT. Documentos, 14).

Wildner, L. do P. and Massignam, A.M. 1994b. Ecofisiologia de alguns adubos verdes de verão: II. Produção de grãos - resultados preliminares. In: *Reunião Centro-sul de adubação verde e rotação de culturas*, 4, 1993, Passo Fundo, RS. Anais. (EMBRAPA - CNPT. *Documentos,* **14**) EMBRAPA-CNPT, Passo Fundo. pp. 142-146.

Wildner, L. do P. and Massignam, A.M. 1994c. Ecofisiologia de alguns adubos verdes de verão: III. Curva de cobertura do solo - resultados preliminares. In: *Reunião Centro-sul de adubação verde e rotação de culturas,* **4**, 1993, Passo Fundo, RS. Anais. EMBRAPA-CNPT, Passo Fundo. pp. 147-150. (EMBRAPA- CNPT. Documentos, 14).

Wilson, G.F., Lal, R. and Okigbo, B.N. 1982. Effects of cover crops on soil structure and on yield of subsequent arable crops grown under strip tillage on an eroded Alfisol. Soil and Tillage Research **2**: 233-250.

Wischmeier and Smith. 1978. Predicting rainfall erosion losses: a guide to conservation planning. Agriculture Handbook, 537. USDA. 58 p.

Wutke, E.B. 1993. Adubação verde: manejo da fitomassa e espécies utilizadas no Estado de São Paulo. In: *Curso sobre Adubação Verde no Instituto Agronômico*, **1**. Wutke, E.B.; Bulisani, E.A. and Mascarenhas, H.A.A. Instituto Agronômico, (Documentos IAC, 35), Campinas. pp. 17-29.

Zenker, R. 1978. Conservação do solo: práticas conservacionistas. (mimeografado). Secretaria da Agricultura, Porto Alegre. 34 p.

Annex 1

Comparison of field work rates with tillage implements

TILLAGE IMPLEMENTS

1. **Conventional tillage**

   "Rome plough" with 16 discs of 26" (66 cm)
   Width 1.65m, working speed 6 km/h
   Work rate = 9.9 ha/day (10 h)

   Light-weight disc harrow with 36 discs of 22" (56 cm)
   Width 3 m, working speed 7 km/h
   Work rate = 21 ha/day (10 h)

2. **Tillage with tined implements**

   Chisel plough with 9 tines:
   Width 3.15m, working speed 6 km/h
   Work rate = 18.9 ha/day (10 h)

   Vertical tined vibro-cultivator with 34 tines:
   Width 3.4m, working speed 10 km/h
   Work rate = 34 ha/day (10 h)

TIME NEEDED TO PREPARE 100 HA WITH A 90 HP TRACTOR

1. CONVENTIONAL TILLAGE

   a. 1 pass with a "Rome Plough" + 2 passes with a light-weight disc harrow
   = 10.10 days + 4.76 days x 2 = 19.6 days

   b. 2 passes with a "Rome Plough" + 2 passes with a light-weight disc harrow
   = 10.10 days x 2 + 4.76 days x 2 = 29.7 days

2. TILLAGE WITH TINED IMPLEMENTS

   a. 1 pass with a chisel plough + 2 passes with a vertical tined vibro-cultivator
   = 5.29 days +2.94 days x 2 = 11.2 days

   b. 2 passes with a chisel plough + 2 passes with a vertical tined vibro-cultivator
   = 5.29 days x 2 + 2.94 days x 2 = 16.5 days

Annex 2

Purchase and maintenance costs for tillage implements

PURCHASE COSTS OF THE IMPLEMENTS (IN US DOLLARS)

1. Conventional tillage implements

"Rome Plough" with 16 discs of 26" (66cm)	4 400
Lightweight disc harrow with 36 discs of 22" (56 cm)	2 200

Total	6 600

2. Implements with vertical tines

Stubble mulch chisel plough with 9 tines and a working width of 3.15m	3 500 (2 500)*
Vertical tined Vibro-cultivator with 34 tines and working width of 3.4m	3 100 (1 700)
Rotary mower with working width of 1.8m	2 000 (2 000)

Total	8 600 (6 200)

*Prices shown in brackets refer to implements fabricated at Santa Cruz.

MAINTENANCE COSTS OF THE IMPLEMENTS OVER 10 YEARS (IN US DOLLARS)

1. Conventional tillage implements

a. "Rome Plough"	
Change all 16 discs every 2 years	3 200
Change all 16 bearings every 10 years	1 000
b. Lightweight disc harrow	
Change all 36 discs every 2 years	2 500
Change all 36 bearings every 10 years	1 300
Lubricants (10 kg/year of grease)	225

Total	8 225

2. **Implements with vertical tines**

a.	Stubble mulch chisel plough Change all 9 points every 2 years	450
b.	Vertical tined Vibro-cultivator (imported) Change all 34 points every 2 years	700
c.	Rotary mower	
	Change the knives twice per year	400
	Change the belts twice per year	400
	Change the oil three times per year	200

	Total	2 150

Note: These costs are calculated over a ten-year period for a farm of 100 ha. The labour charges are not included for changing the discs, bearings and knives, or for lubrication. It is considered that each implement will complete a total of 5 passes per year and that the discs and points will be changed after 1 000 ha of tillage.

SUMMARY OF THE PURCHASE AND TOTAL 10-YEAR MAINTENANCE COSTS FOR THE MAIN TYPES OF TINED AND CONVENTIONAL TILLAGE IMPLEMENTS (IN US DOLLARS)

1. **Conventional tillage implements**

"Rome plough"	8 600
Light-weight disc harrow	6 225

Total	14 825

2. **Implements with vertical tines**

Stubble mulch chisel plough	3 950 (2 950)*
Vertical tined Vibro-cultivator	3 800 (2 400)
Rotary mower	3 000 (3 000)

Total	10 750 (8 350)

* Prices shown in brackets refer to implements fabricated at Santa Cruz.

Annex 3

Notebook for participatory planning in our community

Part 1
Who we are

1. Living in our community are:

(F)1,A	_________ Families
(P)1,B	_________ Persons
(M)1,C	_________ Women
(H)1,D	_________ Men
(+65)1,E	_________ More than 65 years old
(+14)1,F	_________ More than 14 years old
(+7)1,G	_________ More than 7 years old but under 14
(-7)1,H	_________ Under 7 years old

V. Hercilio de Freitas
Brazilian Enterprise for Agricultural Research and Rural Extension
Santa Catarina, Brazil (EPAGRI)

Part 2
Where we are

2. Our community is situated in:

2.1	Continent	
2.2	Country	
2.3	State	
2.4	County	
2.5	Main catchment basin	
2.6	Sub-catchment basin	
2.7	Micro-catchment basin	
2.8	Community	

Part 3
What we have

3. Organization of our community

(A)3.1	Our community has __________ ha of land
(AF)3.2	Each family has an average of __________ ha of land
(D)3.3	The distance of our community from town is __________ km
(E)3.4	Our community has __________ schools
(I)3.5	Our community has __________ churches
(O)3.6	Our community has _____community associations and organizations
(L)3.7	The principal community leaders are: __________________________ __________________________
(AE)3.8	Our community has piped water supplies ( ) Yes ( ) No
(RE)3.9	Our community is connected to the electrical grid system ( ) Yes ( ) No
(T)3.10	Our community has a collective transport system ( ) Yes ( ) No
(TE)3.11	Our community has a telephone ( ) Yes ( ) No

Part 4
What our environmental problems consist of

4. The problems of environmental pollution in our community can be considered

Evaluation grade	Adequate (1)	Tolerable (2)	Worrying (3)	Serious (4)	Very serious (5)
	●	●	●	■	□

Aspects considered	Evaluation	Why?
(L) Domestic refuse	○	
(ES) Industrial waste	○	
(AG) Use of agro-chemicals	○	
(ER) Water erosion	○	
(D) Sewage	○	
Other problems	○	

Part 5
What our socio-economic problems consist of

5. Our socio-economic problems may be considered

Evaluation grade	Adequate (1)	Tolerable (2)	Worrying (3)	Serious (4)	Very serious (5)

Aspects considered	Evaluation	Why?
(S) Health	○	
(AN) Literacy	○	
(AP) Potable water	○	
(L) Electricity	○	
(AL) Food	○	
(M) Housing	○	
(ES) Roads	○	
(T) Transport	○	
(RE) Income	○	
Others	○	

Part 6
What our relations are with nature

6. What is the importance to our community of the natural resources of soil and water?

Evaluation grade	Adequate (1)	Tolerable (2)	Worrying (3)	Serious (4)	Very serious (5)
	●	●	●	■	□

6.1 Soil	Evaluation	Why?
(L) Tillage	○	
(P) Livestock	○	
(F) Forests	○	
Others	○	
6.2 Water		
(CH) Drinking water	○	
(I) Industrial	○	
(IR) Irrigation	○	
(L) Domestic	○	
(CA) For animal consumption	○	
(O) Others	○	

Part 7
Which are the soil management problems in our community?

7. The soil management problems in our community can be considered:

Evaluation grade	Adequate (1)	Tolerable (2)	Worrying (3)	Serious (4)	Very serious (5)
	●	●	●	■	□

Soil	Evaluation	Why?
(F) Fertility	○	
(C) Compaction	○	
(M) Excessive equipment traffic	○	
(FC) Lack of vegetative cover	○	
(Q) Burning of crop residues	○	
(I) Incorporation of crop residues	○	
(E) Soil erosion	○	
(V) Gullies	○	
(NM) Levels of organic matter	○	
(AM) Water storage	○	
(U) Use of the soil beyond its true capability	○	
Others	○	

Participatory action in the plan of execution

LISTING THE ORDER OF PRIORITIES FOR THE ACTIONS

How can we solve the problems of environmental contamination in our community?

1. ______________________________
2. ______________________________
3. ______________________________

How can we solve the socio-economic problems in our community?

1. ______________________________
2. ______________________________
3. ______________________________

How can we solve the soil management problems of our community?

1. ______________________________
2. ______________________________
3. ______________________________

What can assist solving our environmental contamination problems?

1. ____
____
____
____

2. ____
____
____
____

3. ____
____
____
____

When can we solve the problems of environmental contamination?

1. ____
____
____
____

2. ____
____
____
____

3. ____
____
____
____

When can we solve the socio-economic problems of the community?

1. ____
____
____
____

2. ____
____
____
____

3. ____
____
____
____

When can we solve our soil management problems?

1. ____
____
____
____

2. ____
____
____
____

3. ____
____
____
____

WORK PLAN/ OPERATIONAL PLAN/ CALENDAR OF ACTIONS

Community projects

Objectives (What we are going to do)	Methodology (How we are going to do it)	Responsibilities (Who is going to do it)	Timing (When we are going to do it)	Comments

WORK PLAN/ OPERATIONAL PLAN/ CALENDAR OF ACTIONS

Group projects

Objectives (What we are going to do)	Methodology (How we are going to do it)	Responsibilities (Who is going to do it)	Timing (When we are going to do it)	Comments

Thematic maps of the diagnosis

THEMATIC MAP OF THE ENVIRONMENTAL DIAGNOSIS

This map is prepared on the basis of the information placed in the participatory notebook (Part 4), as these refer to environmental problems in the micro-catchment basin.

INDICATOR	SYMBOL
Domestic waste	(L)
Industrial waste	(ES)
Use of agro-chemicals	(AG)
Water erosion	(ER)
Animal waste	(D)
Other problems	(O)

LEGEND

ENVIRONMENT

SYNTHESIS OF THE ENVIRONMENTAL DIAGNOSIS

Micro-catchment basin...................................

Community	(L)	(ES)	(AG)	(ER)	(D)	(O)
01						
02						
03						
04						
05						
Total						

THEMATIC MAP OF THE SOCIO-ECONOMIC DIAGNOSIS

This map is prepared on the basis of the information placed in the participatory notebook (Part 5), as these refer to socio-economic problems.

INDICATOR	SYMBOL
Health	(S)
Education	(E)
Income	(R)
Potable water	(AP)
Electricity	(LU)
Food	(AL)
Housing	(MO)
Roads	(ES)

LEGEND

SOCIOECONOMY

S E R AP LU AL MO ES

SYNTHESIS OF THE SOCIO-ECONOMIC DIAGNOSIS

Micro-catchment basin...

Community	(S)	(E)	(R)	(AP)	(LU)	(AL)	(MO)	(ES)
01								
02								
03								
04								
05								
Total								

THEMATIC MAP OF SOIL MANAGEMENT PROBLEMS

This map is prepared on the basis of the information placed in the participatory notebook (Part 7), as these refer to soil management problems in the micro-catchment basin.

INDICATOR	SYMBOL
Fertility	(F)
Compaction	(C)
Excessive equipment traffic	(M)
Lack of vegetative cover	(FC)
Burning of crop residues	(Q)
Incorporation of crop residues	(I)
Rill erosion	(E)
Gullies	(V)
Organic matter content	(NM)
Water storage	(AM)
Soil use beyond its capability	(U)
Others	(O)

LEGEND

SOIL MANAGEMENT

F C M FC Q I E V NM AM U O

Synthesis of the soil management diagnosis

Micro-catchment basin...

Community	(F)	(C)	(M)	(FC)	(Q)	(I)	(E)	(V)	(NM)	(AM)	(U)	(O)
01												
02												
03												
04												
05												
Total												

FAO TECHNICAL PAPERS

FAO LAND AND WATER BULLETINS

1. Land and water integration and river basin management, 1995 (E)
2. Planning for sustainable use of land resources – Towards a new approach, 1995 (E)
3. Water sector policy review and strategy formulation – A general framework, 1995 (E)
4. Irrigation potential in Africa – A basin approach, 1997 (E)
5. Land quality indicators and their use in sustainable agriculture and rural development, 1997 (E)
6. Long-term scenarios of livestock-crop-land use interactions in developing countries, 1997 (E)
7. Land and water resources information systems, 1998 (E)
8. Manual on integrated soil management and conservation practices, 2000 (E S)

Availability: April 2000

Ar – Arabic
C – Chinese
E – English
F – French
P – Portuguese
S – Spanish

Multil – Multilingual
* Out of print
** In preparation

The FAO Technical Papers are available through the authorized FAO Sales Agents or directly from Sales and Marketing Group, FAO, Viale delle Terme di Caracalla, 00100 Rome, Italy.